BET EARTH
HOW WE CAN STILL WIN THE BIGGEST GAMBLE OF ALL TIME

"Science leaves no doubt that we are in the throes of global warming, but leaves much doubt about how bad it'll get and when. Most importantly, it leaves us puzzling over how much risk of bankrupting the present we should take to reduce the risk of destroying the future. These uncertainties trigger high emotions that have largely polarized policy, political, and public debate; each side predicts looming catastrophe, but for one it's losing the future for investing too little now to save it, while for the other it's losing the present for investing too much. Positing the choice as either/or has stymied the making of major decisions regarding what we can realistically and affordably do, and time is running out. Here comes Professor John Kunich with a simple, plain-English yet highly sophisticated decision matrix for understanding and analyzing these probabilistic questions and hence for finding and accepting suitably proportionate answers. Constructing the matrix from the mathematical theories of Pascal, Gödel, and Heisenberg, Kunich renders the animating ideas completely transparent by colloquially likening the mode of thinking to playing poker the way pros do—coolly, rationally weighing the odds of a hand to know when to hold or fold them. This book is an easy read and, with so much at stake, it's a must one too. Kunich explains why wagering the earth is unavoidable, but why an all-in bet is unthinkable."

DAVID ROSENBERG
Lee S. Kreindler Professor of Law, Harvard Law School

"Kunich's book is a useful addition to thought for environmentalists and for those who distrust environmentalism. Many people claim we are gambling with our world, and Kunich argues that every person needs to assess that claim. His book provides ideas on how we can each make that assessment for ourselves."

RICHARD S. COURTNEY
Expert Peer Reviewer, UN Intergovernmental Panel on Climate Change (IPCC)

"No one understands better than John Kunich—with his grasp of law and environmental issues—the ultimate gamble we face as earth heads for massive disruptions, some certain, some yet preventable."

STUART PIMM
Doris Duke Professor of Conservation Ecology, Duke University

"*Betting the Earth* explores the uneasy parallels between our contemporary environmental challenges and our national fascination with gambling. How much should we bet on preserving biodiversity? Should we bet more on responding to climate change? Where should we place each bet: on federal or state laws, on acquiring public or private preserves, on preventing environmental harms or saving places of special environmental significance? Like it or not, we much make such choices every day, and *Betting the Earth* helps us to understand how we do so."

PROFESSOR JOHN COPELAND NAGLE
John N. Matthews Chair in Law, University of Notre Dame Law School

"I have traveled the world and played in the highest stakes games, or so I thought. Dr. Kunich has shown me that the biggest gamble is all around me and right beneath my feet!"

BARRY GREENSTEIN
high-stakes professional poker player, known as the "Robin Hood of poker"

"Despite my serious doubts about the crusade against global warming, I was extremely impressed by this book. We need objectivity, not the hard sell tactics of so many self-appointed experts. Prof. Kunich helps readers to draw rational, evidence-based conclusions about this vital issue."

ALAN N. SCHOONMAKER, PH.D.
Author of *The Psychology of Poker* and *Poker Winners Are Different*

"In public office, as in private life, we often have to make major decisions based on probabilities and incomplete information. But I can tell you this with great certainty—if you care about environmental decision-making, you should read John Kunich's new book."

CONGRESSMAN ADAM SCHIFF

BETTING THE EARTH
HOW WE CAN STILL WIN THE BIGGEST GAMBLE OF ALL TIME

JOHN CHARLES KUNICH

Parkhurst Brothers, Inc., Publishers
LITTLE ROCK

© Copyright 2010 by John Charles Kunich. All rights reserved. No part of this book may be reproduced in any form, except for brief passages quoted within reviews, without the express prior written consent of Permissions Director, Parkhurst Brothers, Inc., Publishers.

www.pbros.net

Parkhurst Brothers books are distributed to the trade through the Chicago Distribution Center, a unit of the University of Chicago Press, and may be ordered through Ingram Book Company, Baker & Taylor, Follett Library Resources and other book industry wholesalers. To order from the University of Chicago's Chicago Distribution Center, phone 1-800-621-2736 or send a fax to 1-800-621-8476. Copies of this and other Parkhurst Brothers, Inc., Publishers titles are available to organizations and corporations for purchase in quantity by contacting Special Sales Department at our home office location, listed on our website.

Printed in the United States of America

First Edition 2010

2010 2011 2012 2013 2014 2015 2016 2017 2018   18 17 16 15 14 13 12 11 10 9 8 7 6 5 4 3 2 1

Library of Congress Control Number: 2009942591

ISBN: Original Trade Paperback: 978-1-935166-17-7 [10-digit: 1-935166-17-4]
ISBN: e-book: 978-1-935166-28-3 [10-digit: 1-935166-28-X]

This book is printed on archival-quality paper that meets requirements of the American National Standard for Information Sciences, Permanence of Paper, Printed Library Materials, ANSI Z39.48-1984.

Design Director and Dustjacket/cover design:
Wendell E. Hall

Page design:
Shelly Culbertson

Acquired for Parkhurst Brothers, Inc., Publishers by:
Ted Parkhurst

Editor:
Roger Armbrust

Proofreaders:
Bill and Barbara Paddack

DISCOUNTS FOR BULK PURCHASE: Institutions, schools, and organizations may purchase this title in bulk at substantial savings off the suggested list price. Please email randy@pbros.net or refer to our website **www.pbros.net** for current contact information from our Special Sales Department.

Dedication

I humbly dedicate this book to Blaise Pascal, Kurt Gödel, and Werner Heisenberg, three of the greatest minds and spirits ever to grace our beautiful Planet Earth. It is my most sincere hope that my book will make many more people aware of the immense debt we all owe to these Three Wise Men, this trio of incomparable geniuses.

Acknowledgements

I thank my daughters, Christina Laurel Kunich and Julie-Kate Marva Kunich, for their very certain love and support. I know I am always safe when I bet on that. I also thank my wife, Marcia Kathleen Vigil, for her love, patience, and tolerance during my long bouts with this book and my endless ramblings about obscure ideas.

Table of Contents

**11 Chapter 1: High-Stakes Gambling
on Autopilot with the World's Environment**

In this introductory chapter, we discuss current threats to Earth's environment and factors that have brought us to this dangerous point. We examine the political and legislative process that has led to environmental gridlock on the most significant questions our planet has ever faced. We evaluate how leaders have used factual gaps and lack of complete scientific consensus as a reflexive excuse to take no meaningful action, even where the threats seem credible and corrective responses appear both workable and affordable.

**31 Chapter 2: Climate Change:
Doing Something About the Weather**

This features a summary of basic facts and evidence regarding human-generated climate change/global warming. Widely differing collections of data and the resulting divergent conclusions are analyzed and explained. The chapter includes a discussion of ambiguities and uncertainties, as well as the probable consequences and costs of various courses of action and inaction. We examine the bitter partisan conflicts and scandals, including "Climategate," that have distorted the debate and twisted the scientific process into a political battle. We consider the harm caused by bias and political influence, with a view toward finding a rational and fair resolution to the momentous questions at stake.

85 **Chapter 3: Modern Mass Extinction: Twilight of the Living Dead?**

We review the evidence relevant to whether we are now in the midst of the greatest mass extinction since the dinosaurs. We present evidence that enormous numbers of species are currently at risk of extinction, and why it may be rapidly becoming too late to halt this deadly trend. This is contrasted with data showing that very few species are definitely known to have slipped into extinction during the past several centuries. Can both propositions be true? We also discuss the pros and cons of taking effective countermeasures, in light of the uncertain nature and extent of the extinction menace, and the unknowable practical importance of any species threatened with loss.

127 **Chapter 4: Pascal and Decisions in the Midst of the Mist**

We summarize the groundbreaking work of Blaise Pascal, emphasizing Pascal's Wager and his pioneering exploration of probability theory. We discuss how he dealt with important unknowns to arrive at the most rational and wisest decisions on momentous issues. We also analyze how we can apply Pascal's insights and the various forms of decision theory arising from his work to the modern environmental dilemmas we face as a planet.

151 **Chapter 5: Looking for the Missing Pieces of Gödel's Puzzle**

We look at how the genius of Kurt Gödel, and in particular his two incompleteness theorems, can be translated to our global environmental work. On such issues as climate change and a modern mass extinction, his phenomenal mathematical innovations can shed badly-needed light on the proper evaluation of factual incompleteness and inconclusive evidence. We extensively analogize his work on the incompleteness theorems to the ideas in the novel *Flatland* by way of comparison, to add further insight into the problem of solving puzzles with pieces missing.

171 Chapter 6: Heisenberg's Long-range Forecast: Fog Forever

Famous physicist Werner Heisenberg's breakthroughs, including his uncertainty principle, take the spotlight here. We show how his work can illuminate the terrible flaws in our current environmental analysis. Heisenberg can demonstrate that our insistence on scientific consensus is unrealistic and foolhardy. We mention Edward Lorenz's work dealing with chaos theory and the butterfly effect, providing another perspective on the idea of uncertainty and its meaning. The chapter also includes an analysis, guided by Heisenberg's insights, of why so many intelligent people make bad, even self-destructive decisions.

197 Chapter 7: Thinking Inside the Box with the World Wager

Here we discuss the way in which unknowns and scientific questions have been used to block effective environmental legislation and political action. We introduce a better method of incorporating such uncertainty into our environmental decision-making, acknowledging doubts but also providing a prudent risk/benefit calculation. The chapter presents our system for incorporating uncertainty into global environmental decision-making in both the climate change and biodiversity hotspots contexts. We demonstrate how our World Wager and decision matrix take the best features from the work of each intellectual giant previously discussed to formulate a path-breaking tool for the modern age of environmental risk.

231 Chapter 8: You Bet Your Life, and Here's How

This material summarizes the information previously set forth and discusses what's at stake. We consider examples of probability-dependent decision-making from our everyday lives, looking at parallels between them and today's global environmental dilemmas. The focus is on ways we intuitively incorporate uncertainty and probabilities in some of our most important individual decisions, and how we might expand those same approaches to decisions involving the entire planet.

266 **Chapter 9: We're All All-in**

This builds on the previous chapter's analysis, examining some examples of decision-making from gambling activities with which many people are familiar. We consider how we might relate these small-scale gambling experiences to our current worldwide environmental choices. This chapter explores available options in light of all the risks, rewards, and probabilities associated with the dilemma we face. We argue that there is no way to refuse to gamble with the future of Earth; every decision is based on odds and potential loss. That includes a decision to take no action at all and to assume that there is no environmental threat worthy of our attention.

306 **Chapter 10: Now *This* is a Gambling Problem**

This concluding chapter addresses the best way to use the book's new insights to make a real difference for the world's environment. If somehow we can imagine Blaise Pascal, Kurt Gödel, and Werner Heisenberg sitting together around a modern-day poker table, chatting about our most difficult environmental challenges, what might they say? This chapter returns to the World Wager and decision matrix to determine a wise approach to our dilemmas of climate change and mass extinction. It introduces and explains the new concepts of Type R and Type P errors, and uses them to point toward some conclusions. In the highest-stakes game of chance in history, the wisdom of Pascal, Gödel, and Heisenberg, as an extension of their amazing work, could tip the odds in Earth's favor.

358 **Footnotes**

371 **Index**

CHAPTER 1

High-Stakes Gambling on Autopilot with the World's Environment

If you literally had to bet your life on a single choice, with huge risks and costs any way you turn, and with major gaps in the information you need to know, how would you make your decision? Now what would you do if you were forced to bet not just your own life, but also the lives of your family, your friends, and countless millions of other people? How would you decide among the dangerous, costly options, when so many key facts are impossibly beyond your reach?

This isn't a dilemma from a science-fiction novel. It's the terrifying crossroads the whole planet is facing right now, on several global environmental threats. We are truly betting the Earth, today, whether we know it or not. And we can't refuse to gamble. Even if we choose to do nothing at all, that itself is a choice…to bet everything on the proposition that all is well and no changes are justified.

The headlines in 2009 and 2010 were filled with stories about "Climategate" and the furor over false and misleading information related to climate change or global warming. Are powerful people cooking the books and creating the illusion of a monumental peril out of smoke and mirrors? Is the climate change emergency genuine, or an artificial issue concocted to further certain people's hidden agenda?

If climate change is being caused by human activity, and if it will be severe enough to melt ice caps and flood the world's coastal cities, it's a crisis as dire as any we've ever faced. But to have a reasonable chance to halting or reversing that kind of global change would require a cooperative effort of a magnitude and expense the likes of which we've never known. How do we know whether the threat is worth the cure? After "Climategate," what is the reality and what is the illusion? How do we cut through all the politics, hype, spin, and myth and make our decision on a solid foundation of reliable evidence, rational risk/reward analysis, and realistic probabilities?

We're in the same boat, or sinking Ark, on mass extinction. If people are causing vast numbers of species to die out, we could be committing suicide without knowing it. We could be killing off species we've never even discovered that might have one day offered the cure to AIDS, cancer, heart disease, or some horrific epidemic of the future. Human actions might be driving numberless species into extinction, and with them the delicate balance that makes life on Earth possible for all of us. But again, how can we know? If the danger of mass extinction is real and substantial, it would justify major measures to turn the situation around. Is it worth it? Can we fill in all the unknowns, missing pieces, and question marks with enough hard facts to make a reasonable and reasoned decision?

This book tells us how we can make the most important decisions human beings have ever made. And it tells us how we can choose even when we can't know key facts and important variables. We can make incomparably momentous decisions with very incomplete information…and we must. We have no choice. The entire human race is gambling, pushing all-in, and betting everything we have on these Earth-shaking issues, with holes in the facts everywhere we look. This book shows how we can make the uncertainties, and the gaps in our information, work for us rather than against us.

We're playing for the highest stakes any gambler has ever

imagined—and we'd better be right. I will explain how some of the greatest minds in history can help us find our way through the fog and the darkness to make the correct choices. This is no time for politics, cooked books, easy assumptions, or smug complacency. All of our chips, and all of our lives are at stake. We need more than luck on our side at a time like this. That's why this book was written.

It's human nature to want to back a winner. People like to bet on a sure thing. As poker players know, when you push all-in, you'd like to have "the nuts," the best possible cards. Doubt is disturbing to many individuals, especially when the fog of uncertainty shrouds some of our most weighty personal decisions. When we need to decide which school to attend, what career to pursue, whom to choose as a life partner, what house to buy, how to resolve major ethical or moral problems, whether to have children, or what job offer to accept, we can agonize greatly, not knowing how to sort out all the variables. Maybe you've never encountered such dilemmas, but I have. I am an official, permanent legal resident of the great State of Confusion.

Let's be honest about it. We all have experience wrestling with such doubt-drenched, life-changing decision points in our own lives. We all know it's difficult and perplexing to be compelled to decide Big Issues despite immense gaps in our information foundation. When we are forced to make a choice that will forever change the direction of our lives, we want to know all the facts, in advance, right up front. We want to leave nothing in doubt, and trust nothing to chance with so much at stake. It seems only fair, right? Instead, life has an annoying habit of blindfolding us, spinning us around like a top, and then airlifting us right onto a midnight freight car aboard the speeding Disoriented Express.

This book takes that painful personal-crisis situation and transposes it to the highest levels of human endeavor. We're going to take all that individual angst and agony and multiply it by a factor of several billion—not because that's fun or pleasant, but because it's unimaginably important. We must find the right answers to world-changing questions today, at the crossroads of Earth's entire history. When national and international decision-makers stare into a future—or even a present day—that is at best dimly illuminated, how can they resolve some of the most momentous turning-point hazards that have ever threatened this planet?

I will examine in detail just a couple of the greatest challenges of this or any age, in the field of international environmental policy and law. The full list would include daunting global scientific, technical, and legal issues involving biodiversity, natural resources, climate change, energy policy, nuclear power, human population growth and decline, habitat destruction, and transboundary pollution, plus many more. But using two prime examples, I will construct a paradigm for handling all of these immense problems in light of the persistent, gigantic holes in our knowledge base. To do so, it will be necessary to draw on pathbreaking work from three of the most brilliant minds ever to grace the earth's surface. Blaise Pascal, Kurt Gödel, and Werner Heisenberg can provide a way for us to navigate through the thickest fog in our search for the right path, even while the most horrific perils lie in wait for us to make the wrong move. If you haven't heard of these men, or if you don't know much about them, keep reading. I hope to make them come alive for you and prove how vital their work is to all of us today.

This is a matter of tremendous importance, because the wrong approach to scientific or factual uncertainty can cause us to opt for inaction where action is desperately needed, or to squander immense resources on illusory or insignificant mirages. Leaders today often possess a natural, widespread, and badly misguided penchant to select "do nothing" as our default option when environmental issues come bundled with layers and layers of doubt. I intend to show that—in the area of uncertainty—the finest and most original thinking of all history may at times point in exactly the opposite direction from the reflexive impulse to stick with the stand-pat status quo. *The key is to be able to recognize this situation when it comes up.* This book will demonstrate how we should formulate crucial environmental decisions in the dim light of unknowable facts and gap-riddled evidence, and how we can discern when action rather than static cling is mandated...and vice versa. A world of difference exists between "status quo" and "let's go" circumstances, and we've got to be able to tell the two apart.

When politicians must decide whether to support substantial new environmental measures, many have a ready-made self-inking rubber stamp handy, bearing the single word, "No!", complete with exclamation point and large, easy-to-read lettering. This is true in nation after nation, all over the world. The default option is almost always to refuse any efforts to impose expensive and burdensome new environmental restrictions, regulations, and rules. These "three Rs" are easy for political leaders to ignore, largely because they usually lack a large, politically powerful, and highly motivated constituency fighting hard for them.

Environmental issues often fail to ignite a massive committed base of political grass-roots support because they lack obvious, apparent impact and immediacy. Citizens become outraged and energized over issues that they perceive as having a direct, tangible, physical, visible, and near-term effect on themselves and their families. Voters, and the politicians who depend on their support at the ballot box, pay attention to reports of a local factory shutting down, plummeting property values, rising gasoline prices, nearby terrorist attacks, massive layoffs, skyrocketing health care costs, mortgage problems, violent local crime rates, and escalating home-town casualties from the latest war. This type of phenomenon makes an unmistakable and real-time mark on people. It hits us where we live, literally, and it hits us where it hurts. Virtually everyone will either be affected by these issues personally or know someone who is. Everyone is capable of verifying the reality of such hot-button topics without any need to trust the word of experts or to take the proverbial leap of faith. No faith-leaping is required when you can see and feel the incontrovertible, painful evidence for yourself, just beyond your living room window, or even smack in the middle of your own living room, right in the here and now.

Politicians respond to voter pressure, and voter pressure usually builds up to action-provoking levels only when the typical layperson perceives a genuine, practical, imminent threat to her—or her family's—own health or wealth. Because of this, governments tend to be reactive, not proactive. Leaders are often actually following the lead of their constituencies, not the other way around. When a politician insists on awaiting the results of a local focus group and a few carefully targeted polls before he or she is willing to formulate a policy position, no one should be shocked when the "dog" of political action is often wagged by the "tail" of popular concern and opinion, whether or not it is well-rooted in fact and logic.

This wouldn't be a particularly serious problem if the *vox populi* prudently focused on issues of legitimate gravity, including long-term, slow-acting threats, and embracing both local and very non-local challenges. Unfortunately, many of us prefer to stay out of political controversies unless they hit us between the eyes. Hard. Our field of vision is narrow, and we are decidedly nearsighted. Influences that do not have a conspicuous impact on our own locality, and/or are projected to happen more than a few years off in Tomorrowland, are much more likely to escape our notice than those taking place right now in our own front yard. This major factor contributes to the absence of congressional action regarding inadequacies in the Social Security system, to cite just one well-known example. If drastic tax increases and/or deep cuts in benefits and services are not expected to take place within our own lifetime, if at all, the natural response of most voters is to relax and whistle another chorus of the old favorite song, "Don't Worry, Be Happy."

In no other area is our collective political myopia more severe than the category of environmental protection. If an environmental threat primarily touches on other parts of the world, it's as if it were on a different planet in another solar system in a remote galaxy many light years away. And if an environmental problem is not causing dramatic ill effects that we can see with our own eyes right now, we behave as if there were no problem whatsoever. For many of us, without a headline-grabbing current crisis that we perceive as imperiling ourselves at home here today, environmental regulation holds a stand-by ticket to take a back seat on the last bus in our long convoy of political issues. At best.

Do you want evidence of our environmental near-sightedness? Imagine yourself at an eye examination, looking at a very simple eye chart. The chart has only two lines of text. The top line consists of a string of letters that spell out ENVIRONMENTAL HAZARD IN THIS TOWN, while the bottom line reads ENVIRONMENTAL HAZARD IN SOUTHEAST ASIA. Which line do you think you would find easier to see? Now switch to a second eye chart as we continue to play our pick-up game of "Is this better, or is that better?" This chart again has only two lines. The top one says DANGER NOW, and the bottom row spells DANGER IN MAYBE FIVE HUNDRED YEARS. Is the top one clearer, or the bottom one? Would you like UV protection with your new glasses?

The widespread preference for taking care of problems we see as immediate and close to home is very natural and understandable. It's part of our survival instinct. We're hard-wired to react much more strongly to threats that are close to us and our loved ones at the moment than to those we can't even perceive because they are many miles away, or off in the distant future, or both. This type of local-focal backyard immediacy is crucial to self-preservation in the wild where imminent danger to ourselves and our kids often lurks behind every tree. But it doesn't translate well to our modern, globalized, interdependent way of life where deadly threats can just as easily originate from half a world away as from behind the next boulder. Until very recently, and I mean within the past 50 years or so, it made very good sense for most people to take the Vito Corleone attitude, "I'm not interested in things that don't concern me." In the not-so-long-ago era depicted in *The Godfather Part II*, we could be confident that we knew what things concerned us, because they were very nearby and likely to happen soon enough for us to see them coming. It isn't at all easy to unlearn such time-tested survival mechanisms that consistently and reliably pulled our ancestor's asses out of the reach of countless ravenous jaws for millions of years of trial-and-error living. In fact, to this day many of us haven't even felt any particular need to fix something that doesn't show any obvious signs of being broken. Like a comfortable old pair of slippers, we keep sticking our feet into the same up-close-and-personal radar scope that worked so well for everyone from the post-dinosaur-dust-up, clean-up crew to the Godfather.

The United States' history of federal environmental law and policy provides ample evidence of our predilection to kick the can far down the road unless the can is completely rusted through and dripping burning acid on our big toe at this very moment. Several of the major environmental statutes became law only when an issue suddenly became mass-media, front-burner material, and Congress felt intense political pressure to do something, anything, to show the public they were on top of the matter. The underlying problems may have been spreading and deepening for dozens or hundreds of years, steadily going from bad to worse to dreadful; but without that high-profile photo-op, we had no answer when someone asked, "What's in it for me?"

For example, the Endangered Species Act[1] emerged from the outcry generated by Rachel Carson's[2] book, *Silent Spring*[3], and the danger that the American Bald Eagle might become extinct due to the harmful effects of such pesticides as DDT. Similarly, the Love Canal hazardous waste controversy[4] helped launch the Comprehensive Environmental Response, Compensation, and Liability Act (CERCLA)[5] and its centerpiece, Superfund. The Toxic Substances Control Act (TSCA)[6] became the vehicle for addressing in a highly-visible way the widespread public concern generated by well-publicized scares regarding asbestos,[7] PCBs,[8] and chemicals believed to harm Earth's protective ozone layer.[9] And the much-discussed risks from acid rain[10] sparked sufficient furor among the citizenry[11] that Congress added an entire new section to the Clean Air Act in 1990 to respond.[12]

Each of these legislative measures, and several others like them, were appropriate and useful responses to genuine environmental issues. But in all likelihood they never would have been seriously debated by the United States Congress, let alone enacted into law, but for the serendipitous emergence of a high-visibility focal point that attracted vigorous public attention and generated strong pressure for legislative action. Hundreds of badly decimated species were in great need of federal protection in the United States long before the Bald Eagle/DDT issue spiked up, but none of them was our national symbol, and none of them was the core subject of a famous book.[13] We needed to regulate toxic waste long before anyone knew the difference between Love Canal and *The Love Boat*, but it wasn't making it into headlines all over the nation. Many important environmental statutes owe their existence to the happenstance of an imagination-capturing catalyst that shocked an inattentive and apathetic populace into action mode. Conversely, the issues that have not been "blessed" with a high-profile mass media frenzy have often gone unaddressed by legislators, regardless of whether they are in actuality as significant as—or even more significant than—those in the spotlight.

The economic catastrophes that erupted in late 2008 have provided us with some fresh examples of this crisis focus in real-world application. As the newly-installed Obama Administration sought to enact or implement major portions of their agenda, some tied their proposals to the widely perceived urgent need to rectify the collapsing economy. The idea: Use the strong popular and political support in favor of emergency corrective action on the recession to piggyback a host of legislative measures across the finish line, where this would be practically impossible under non-crisis circumstances. White House Chief of Staff Rahm Emmanuel was quoted as saying, "You don't ever want a crisis to go to waste; it's an opportunity to do important things that you would otherwise avoid."[14] Secretary of State Hilary Clinton similarly stated, "Never waste a good crisis....Don't waste it when it can have a very positive impact on climate change and energy security."[15] They were speaking as experienced politicians who understand that politics truly is the "art of the possible," and that the realm of the possible expands and contracts in unison with our perception of the urgency and immediacy of the situation. In other words, nothing lights a fire under a politician like lighting a fire under a politician.

This highlights the deadliest danger inherent in environmental protection today. Unless political leaders and a sizable, motivated segment of the electorate *perceive* an environmental situation as a problem worthy of their attention, they will do nothing. Here is a sad, sobering truth: In terms of spurring legislation, it is politically irrelevant that a particular phenomenon might in fact be a mortal threat to the entire planet…if a sufficient number of the citizens and politicians don't believe it. We're proving with all the evidence in the world that the cliché is true: Seeing is believing, and that which we can't see, or feel the sting from, might as well be a dodo's pet unicorn for all the weight we give it in our choice-making.

A rather widely-held attitude out there views any environmental issue as falling on the priority scale somewhere between "maybe later" and "fuhgeddaboutit." This viewpoint applies across the entire spectrum of environmental concerns, even to the most well-documented, indisputable, thoroughly proven hazards. People who subscribe to this mindset sometimes see all environmentalists and their pet causes as frivolous, anti-progress, save-the-owls relics from a hippie commune's tie-dyed time capsule. They tend to presume that environmental causes are rarely if ever as dire as the activists claim them to be, but rather are nothing more than trumped-up cases of zealots crying "endangered wolf." Skeptics are often jaded by what they see as a long series of panicky news reports and frantic medical studies unnecessarily scaring people with urban myths about poisoned apples in their supermarket and radioactive Godzillas with anger-management issues emerging from the glowing rubble of Three Mile Island.

It is an inescapable fact of life that environmental threats, like most health hazards, usually do their dirty work slowly, inconspicuously, and with plenty of camouflage. Whether the danger involves cigarette smoking, obesity, or elevated levels of water pollution, the damage is done little by little over a long period of time, working alongside many other factors that also exert some impact on our life and health. Unlike an immediate, dramatic assault in the form of a gunshot, these stressors operate at the level of slight changes in probability, sustained over prolonged experience, with no obvious negative effects or tangible symptoms for extended stretches of apparent nothingness. Because we often don't witness any alarming evidence of trouble for years and years of exposure to a given environmental or health threat, it's very easy for people predisposed to a dubious mindset to assume that a whole lot of nothing is going on. If seeing is believing, how can we believe in a crisis that is so persistently invisible? It's no contest when an environmental issue has to compete for public attention and governmental intervention with palpable, immediate, kicking-in-the-front-door, hot-button headlines like a devastating recession or a deadly crime wave.

If environmental dangers start the law-and-policy race a couple of laps behind—even when they are firmly supported by unanimous and abundant evidence—imagine what happens when we add in heaping loads of scientific dispute, inconclusive evidence, theoretical disagreement, and speculative cause-and-effect relationships. That's right—what usually happens is nothing. Call it what you will—uncertainty, incomplete information, unknowns, questionable causation, unclear probabilities—once you factor in a significant element of doubt, environmental causes become lucky to make it into the kitchen at all, let alone anywhere as close to being hot as warming up on the back burner.

Tragically, people fail to believe in some real environmental hazards because they view *any* uncertainty as fatal to the cause of regulation. The prevailing attitude is that the best choice is to stick to the status quo until and unless the pertinent scientific and factual question marks have been definitively resolved. This mindset insists that we maintain the default option of no action in the face of even minor or remote, but seemingly legitimate, doubts within the scientific community. And when, as is often the case in the environmental science field, many people are poorly prepared to identify and assess complex issues for themselves, it's common for laypersons to defer to the opinions of those they see as experts. Even where the self-promoting experts really aren't much more capable of forming a reasonable view than our next-door neighbor, it's intimidating even to think about taking a stand in opposition to big-name public figures, so we tend to have consensus by default. Thus, if respectable and well-publicized expert opinion appears to be in agreement, we usually go along with the crowd. But if there are glittering credentials on both sides of the issue, the existence of doubt spells an early end for most environmental initiatives. When the experts rumble and grumble amongst themselves, who are we to be the referee? A person could get hurt in there. It's a lot less sweaty and bloody on the outside of the boxing ring.

People are very reluctant to cast their votes and expend their money and effort on environmental issues when they are unsure a real problem exists. Why risk all that time, money, and effort on a cause that may be a mirage? There is never a shortage of no-kidding, in-your-face, no-doubt-about-it, undeniably real challenges and emergencies right in front of our eyes, demanding our prompt attention. All those home-turf problems are always there, and we don't have to go searching for them; we trip over them or step into them all the time. In fact, they all too often find *us* and make their presence felt as undeniably as a baseball bat to the forehead. It is quite a hard sell to persuade people to turn their focus away from this horrifying horde of home-town horribles even for a part-time bout with an uncertain threat. And this built-in inertia is exacerbated when people view a problem as distant, way off in the future, far from home, not directly threatening to humans, foreign, or in any way not within their own immediate bubble. An environmental issue burdened with a combination of physical and temporal remoteness and factual doubt evokes at an absolute maximum only one response from most people: a shrug. Or as my young daughter Christie would so eloquently yet succinctly put it, "Whatever."

Because of this systemic bias against environmentally protective action, two opposing stereotypes have arisen. The over-generalized archetypes are cartoonish, as stereotypes usually are, but they contribute to this distortion by regularly depicting their opponents in highly exaggerated form. On one side there are true-green environmentalists who portray issues of concern to them in the most extreme terms imaginable. They over-state the evidence in support of the threats, and caricaturize skeptics as "denialists" on a par with flat-Earth devotees or Holocaust deniers…their opponents are either wack-jobs or whackers. They view any opposition as misguided at best, and evil at worst. The other stereotype consists of the anti-environment, pro-development robber-barons whose craving for profits leads them to dismiss every environmental hazard as "alarmist" or a hoax. The only green things they value are greenbacks. They dismiss environmentalists as sky-is-falling, anti-capitalist radicals who only want to stifle economic growth, usher in a socialist government, phase out all human beings, and (in the interim) give group-hugs to trees.

A polarized, oppositional paradigm is a natural consequence of our profound, widespread misunderstanding of scientific and factual uncertainty. In our ignorance, many of us have gravitated to the primitive rallying cry, "When in doubt, fight it out!" And when people feel compelled to take sides, as if in battle, it is entirely predictable that mutual cooperation will be at a minimum and hard-wired gridlock will be the norm. Legitimate and open-minded communication across political aisles that more closely resemble battle lines will be a rarity, and genuine efforts to understand and appreciate differing points of view will be an anomaly, if not evidence of heresy or treason. It is almost akin to the great religious wars, where each belligerent camp fervently believes in the apodictic and exclusive truthfulness of its own faith, and anathematizes its opponents as damnable (or already damned) heretics and infidels. When any cause takes on the trappings of a mystical faith, complete with sacred dogma, zealous adherents, and a vast us-versus-them, right-versus-wrong divide between believers and skeptics, the chances of real cooperation crash down hard. It is a poisonous recipe for conflict, combativeness, and propaganda...anything but effective and appropriate action.

Before you start lighting torches and grabbing pitchforks to chase me to the nearest cliff, let me emphasize that I have nothing against religion, or religious faith. They can be beautiful and inspirational sources of personal joy that establish the foundation for a fulfilling life. But outside of a theocracy, reliance on religious doctrine or an equivalent secular corollary is a poor way of formulating official governmental policy and law. If we want our collective actions to be consistent with the best available evidence and realistic, rational thought on risks and rewards, advantages and dangers, costs and benefits, we won't find much support in a system that, by definition, exists independent of evidence and reason. Faith is a committed belief in something we can't prove or verify through the prosaic means of fact-finding, logic, objective observation, and scientific experimentation. This doesn't mean that faith is irrational, crazy, contrary to all evidence, or in opposition to reality...only that it can—and is meant to—operate without the foundation of cold, hard, rigorous proof we demand in a court of law.

What are some distinctive characteristics we might associate with a mode of decision-making that is rooted in faith? I think a faith-based system (whether religious or secular) would be based on beliefs that are not definitively provable through empirical observation or experiment. It holds positions, and often holds them very strongly, even where there is little or no objective and verifiable evidence in their support. It demands, and many times receives, the whole-hearted commitment of a cadre of devoted adherents whose lives are largely focused on the core principles of the faith. A faith-based paradigm generally is very resistant to challenges from other faiths, or from argument and evidence derived from such foreign disciplines as science, history, sociology, logic, or mathematics. People believe what they believe, and can be quite steadfast even in the face of what others might view as powerful and obvious contrary facts and principles. Faithful devotees tend to band together with other true believers in the same core nucleus of doctrines and dogmas, and to separate themselves from those with beliefs they view as heretical, evil, stupid, unenlightened, ignorant, blasphemous, apostate, or otherwise in opposition to incontrovertibly revealed Truth.

This insider/outsider aspect of a faith-rooted belief system springs from the idea that the faith is absolutely true and good, irrespective of oppositional assertions or evidence. If the central tenets that drive our major life choices are indisputably true, noble, and of paramount importance, then anything that threatens them or differs with them will often be seen as dangerous and malevolent. Absolute truth, being highly resistant to attack by logic, fact, scientific theory, or contrary belief, will tend to inspire fierce loyalty from its followers, who will think of those who dispute and challenge that truth as infidels, traitors, devils, heathens, or deluded fools. Unbelievers, meaning the people in such benighted and fallen condition, are beyond the realm of the saved—they are to be excommunicated, anathematized, shunned, demonized, ridiculed, dehumanized, and even fought. And devotees will often view outsiders who don't directly disavow the truth as potential converts to the true faith, waiting to be enlightened and proselytized by the faithful missionaries among the insiders.

In later chapters, I'll return to this notion when I discuss in detail the role of uncertainty and probability in decision-making. I will look at ways in which some of our most vital environmental issues have been handled in a manner more resembling a faith-based religious doctrinal dispute than an evidence-rooted phenomenological problem. At this point I just want to identify this as an obstacle to making wise choices.

What's wrong with building opinions on faith as a springboard for public policy and legislative action? I see the primary difficulty as one of unproductive debate and ineffective communication. How do you usually react when missionaries or representatives of a religion to which you don't belong come uninvited to your door and try to persuade you to convert to their faith? What is your reaction when friends, relatives, or co-workers of differing faiths repeatedly bring up their religious views and attempt to sway you to their position? How often do these discussions result in anyone switching to a new belief system, or moving significantly in that direction? What lines of persuasion do people typically use during these exchanges? What types of evidence or appeals to authority generally come up? How do you feel when you become party to one of these proselytizing discussions? Are such conversations often emotional, uncomfortable, or unpleasant, or do you look forward to them and actively seek out opportunities for more of them?

In my experience, any variety of missionary discussion often entails an appeal to what is believed to be indisputable, divine authority in the form of sacred writings or holy scripture (such as the Bible, Quran, or Book of Mormon). People frequently urge others to meditate and pray about the matter, to ask God what course of action is right. In trying to convert others, people may share examples from their own personal experiences in which their lives have been improved through adherence to a particular religious faith. These testimonies are often deeply heartfelt and profoundly emotional, and people who relate them to potential converts may shed tears of genuine joy and core conviction as they talk.

All of these features are natural and appropriate in the context of religious faith, which is intrinsically on a different plane from the physical world of facts, numbers, science, and history. But when we try to graft a faith-based mode of reasoning onto a legal or policy problem, it's a poor fit. Who knew? It's exceedingly difficult to have meaningful and open debate leading to positive action when the participants insist on acceptance of a set of principles and supposed facts regardless of contrary evidence or opposing argument. A doctrinal dispute rooted in faith leaves little or no room for objective analysis of all the evidence or fair appraisal of all competing lines of reasoning. When entire catalogues of evidence, loaded libraries' worth of ideas, and whole bodies of thought are ruled out of bounds categorically and automatically, the opportunities for real give-and-take collaboration are, at best, slim. If we summarily exclude entire categories of data and sources of information as inherently unreliable, we transform the debate from one of logical rationality into something like a religious argument. We still might arrive at a reasonable final decision, but the odds of doing so go way down once we preemptively eliminate much of the available relevant evidence. It's like trying to piece together a jigsaw puzzle with most of the lights turned off. Interesting, maybe, but needlessly difficult.

Within the environmental context, many people may have instinctively gravitated to a quasi-religious analytical framework because they see some of the same key factors at work in both the environmental and spiritual realms. Arguably, they have several elements in common, including: (1) a concern with matters of enormous significance; (2) the need to make momentous, future-shaping choices; (3) strong, persistent disagreement on the most appropriate course of action; (4) the existence of some evidence that can plausibly be interpreted as tending to support two or more opposing conclusions; and (5) a considerable degree of uncertainty and lack of objectively definitive evidence on some major points relevant to the decisions at hand. While some take a split of opinion and absence of incontrovertible proof as ample justification to stand pat, others react with a faith-based response. They form an opinion based on feelings, emotions, personal experience, peer attitudes, social norms, and general predispositions, buttressed with a selective sampling of the available evidence, and take it on faith that their choice is correct.

The quasi-religious approach to environmental issues doesn't play favorites. There are hosts of true believers (and true unbelievers) on both sides of crucial subjects that include biodiversity loss, the ozone layer, nuclear power, and climate change. Content with a judgment that feels right to them and is a comfortable fit for them in their social network, they consider the matter closed and fiercely resist any attempts to prove them wrong. You can find throngs of thoroughly faithful, orthodox devotees of both environmental extremes, from "All is well" to "All is Hell," united in their unwillingness to consider any facts that don't jibe with their chosen dogma. Defenders of the True Faith are always zealous, fervent, deeply committed, loyal, and steadfast in combating their religion's perceived enemies, and both of the environmental varieties are right in line and standing at attention. As a mechanism for rationally and realistically resolving complex, immensely important questions, though, all that fervid preaching can get in the way of clear thinking and cooperative analysis.

This perverse misconception of uncertainty is exactly what has brought about some of the worst global environmental perils in our planet's history. Despite the presence of huge governmental bureaucracies all over the world on the national, regional, and international level, the legal response to these threats has run the gamut all the way from nothing to half-hearted. I wrote this book to uncover the real reasons why the collective legal effort on such immense challenges has been so woefully inadequate. A brief look at just two of the worst, yet unchecked, problems may begin to demonstrate why the law has done so little about so much for so long. There are other issues that have gone without an efficacious legal response for much the same reasons—including ozone layer depletion, transboundary pollution, consequences of choosing fossil fuels versus nuclear power, the proper role of genetically modified organisms, and the political/economic impact on environmental quality—but I will consider just two as representative of a much broader syndrome. They are climate change and mass extinction, which I will now consider in that order in the next two chapters.

Keep in mind that they are here to serve as illustrative examples. They're detailed and extremely important examples, but still only examples, selected from among numerous other possibilities. I'm going to dig into them in enough depth to show you how unknowns, uncertainties, variables, scientific disagreements, and missing factual pieces have been used—and misused—to generate a whole lot of nothing. Now sometimes "nothing" is what a particular situation really calls for. Not every postulate is a problem in disguise, needing corrective action. But sometimes we legitimately find ourselves in a situation that demands intervention, even when we are beset by swarms of annoying doubts and gaps buzzing all around us like hazy, hard-to-swat mosquitoes.

A bit later in this book I'm going to rely on Pascal, Gödel, Heisenberg, and other great intellects to help us find our way through the fog of uncertain and incomplete information. I will make my case that, whether we know it or not, we are currently betting the Earth with some of the choices we're making—and we're often making those choices in a dark room with three blindfolds on. And I'll discuss my own ideas for a rational, logical, objective, and balanced approach to decision-making that incorporates and makes adjustments for gaps in the evidence and questions about the nature of the reality we face. But that portion of the book won't have an anchor to real-world issues unless we first spend some time examining at least two of the most momentous environmental challenges of all time. So let's see where we are, how we got there, and why we are in such desperate need of help.

CHAPTER 2

Climate Change: Doing Something About the Weather

Climate change, sometimes referred to as global warming, has been a matter of great controversy in recent years. Many entire books are devoted to this complex and difficult issue, and I don't intend to work up a sweat (sorry) by foolishly attempting to cover it comprehensively in this one chapter, however brilliant it might be. Luckily, I don't need to, because the ultimate purpose of this book doesn't hinge on the precise nature and degree—or even the reality—of the climate change threat, or on what specific steps might be warranted in response. *This is a book about how we can tell whether we have a problem worth our considerable trouble to correct it, when significant relevant factors remain beyond our ability to resolve.* Given that perspective, I will limit this discussion to a brief big-picture overview of just some of the evidence for and against the propositions that (1) climate change poses a serious danger and will continue to do so; (2) human beings are causing and/or might be able to prevent major damage through some form of intervention; and (3) any attempt at corrective action could be ineffective, excessively expensive, and/or demanding.

It is most definitely *not* my intention to set forth in anything approaching a comprehensive treatment the voluminous, frequently ambiguous, and often highly technical scientific data tending to establish the dimensions, origins, and potential corrective actions relative to any current climate-change crisis. Climate change on a global scale is an issue replete with innumerable charts, graphs, tables of numbers, trend lines, and impressive-looking diagrams. And you can assemble a daunting set of such figures on the side of both the reality and the illusion of a modern climatic catastrophe-in-progress. The point of this book doesn't turn on which is the correct view, so I won't pretend to resolve that dispute here. Neither will I propose a workable, let alone an ideal, solution to whatever hazard we now face from global warming.

So what if anything will I actually do in this chapter? I'll merely try to satisfy you that there is *some* credible scientific evidence on both sides of the climate change issue, and that our ability to erase any relevant question marks any time soon is itself questionable at best. That's it. That is all I really need to do to set the stage for my proposed pathway out of the fog and smoke. So let's get to it.

What is at the heart of the climate change issue? The problem centers on the heat-trapping properties of so-called "greenhouse gasses." Greenhouse gasses are hydrocarbons (comprised of various combinations of, primarily, carbon and hydrogen), and include such familiar and natural chemical substances as carbon dioxide, methane, and similar gasses. For more than a century, there has been scientifically sound evidence that an atmosphere rich in such gasses acts to hold more heat in, close to Earth's surface, rather than allow it to radiate outward into space as readily as it would in an atmosphere containing less of these gasses. By interfering with the transfer of infrared energy in the atmosphere, greenhouse gasses can cause Earth's surface, including both land and water, to become generally warmer, all else being equal. That's the basic idea at its most simplistic level, neglecting the complex interplay of multiple processes and feedback mechanisms.

Planet-wide emissions of greenhouse gasses such as carbon dioxide and methane have increased dramatically since the dawn of the industrial age. Automobile exhaust, factory emissions, methane from domesticated livestock, fugitive emissions from gas stations, and other sources have soared as populations have grown and human activity has shifted toward pursuits that entail a net gain of greenhouse gasses in the atmosphere. Simultaneously, humans have drastically reduced the amount of land covered by forests and other vegetation, thereby eliminating the potential for enormous amounts of carbon dioxide to be removed from the atmosphere and converted into oxygen by all those photosynthesizing trees and other plants. There are also some indications that the oceans are now absorbing less carbon dioxide than before, reducing the effectiveness of another important means of lowering the amount in the air. The combination of higher emissions and lower photosynthesis/absorption has caused a larger quantity of greenhouse gasses to enter and remain in the atmosphere…a big net increase in the amount of hydrocarbons in the air on planet Earth.

Scientists are able to gain indirect information as to the composition of Earth's atmosphere at various points in the long-ago past by analyzing samples trapped in ice cores, taken from ancient glaciers and polar ice caps. There are significant long-term ice deposits in such locations as Antarctica, Greenland, and tropical mountain peaks. By sampling these very old, icy time capsules at differing depths, researchers can gauge what our planet's atmospheric composition was back in the very-distant day when such natural freezer-packs were first formed. We (and by "we" I, of course, mean some brave and hardy people other than myself, because I opt to remain ever warm and comfortable) now have been able to reach an incredible two miles deep into the Antarctic ice and retrieve minuscule gas samples from air bubbles locked in place as long ago as 650 millennia. You might think that anything that old would be uncontroversial, but in fact almost nothing regarding climate change is uncontroversial...even the most fossilized varieties of ancient history.

It requires the skill and knowledge of a climatological version of Sherlock Holmes to examine subtle clues from tree rings, corals, and tiny gas samples from long, long ago and make reasonably reliable inferences as to what they signify in terms of our planet's overall temperature back then. Have you ever tried to deduce anything about the Earth's climate from ice core samples millions of years old? If that isn't the ultimate "Cold Case," I'd like to see what is. With perhaps a greater degree of understatement than I typically employ, let me say that conclusions extracted from such climatic detective work is open to varying interpretations. But given that such evidence is all we have once we look to any period earlier than the Industrial Revolution, we have to take what we can get, and remember to take it with a grain of salt or two. Much depends on our analysis of those ancient, very cold, cases when we're trying to put more recent trends into perspective, so this is an enterprise of great importance as well as significant uncertainty.

For example, what do we see when we make a graph showing long-term trends in global (or at least large area) mean temperature? It depends very much on who is making the graph and what data they include. For data from more than two or three centuries ago, a lot hinges on what evidence we emphasize and what regions we focus on. Even for very modern data, it can matter whether we concentrate on surface readings or evidence from satellites, and what methods we use to derive our overall data points. When we're playing "connect the dots" to decide whether we're in a climate-change crisis, it is fundamental that we pick valid dots before we connect them. Let's look at some examples of the ways our planet's climatic trends vary according to the eye of the beholder.

When Michael Mann and other prominent scientists compared the results from a selection of older samples (from tree rings, corals, and ice samples) with the direct measurements that have been possible during contemporary times, they found something alarming. The resulting graph consisted mostly of a long, essentially flat EKG-like line with smallish peaks and valleys for century after century...until you come to a point somewhere around the year 1900. Since then, the trend in average temperature readings, at least for the Northern Hemisphere, has spiked at a steep upward angle, rocketing vertically to a level that had never been reached for at least two-thirds of a million years. This is known as the "hockey stick" graph because it's mostly long and straight for hundreds or even thousands of years, with a short upward angle at the tip, reflecting the post-1900 temperature climb.

The hockey stick graph was first published in 1998,[1] and it was featured in the 2001 Third Assessment Report of the United Nations Intergovernmental Panel on Climate Change (IPCC),[2] where it quickly received a great deal of press attention. The graph was also included in the 2007 IPCC report.[3] It dramatically represented the apparently unprecedented surge in Earth's temperature during the most recent years. The steep slope of the upward end of the stick, if continued, seemed clearly to warn of much more significant and potentially ruinous temperature increases in the years just ahead.

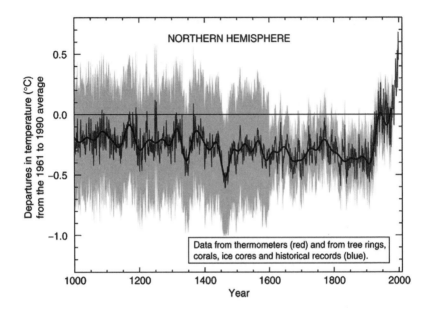

According to National Oceanic and Atmospheric Agency (NOAA) statistics, the concentration of atmospheric carbon dioxide has surged as well, going in the same direction as temperature, as would be predicted by the greenhouse effect. Only during the last few decades has the amount of carbon dioxide ever exceeded 300 parts per million in Earth's atmosphere. Clearly, for the many scientists who consider the hockey stick graph to be valid, the meaning is significant: Since the advent of the Industrial Revolution, a new factor has been incorporated into the global climate equation, and we are the ones "doing the math."

However, some scientists have disputed the accuracy of the hockey stick graph.[4] They argue that it is misleading, and that there is no real evidence of a recent unprecedented spike in global temperatures. One of their points is that the hockey stick graph erroneously omits evidence of the Medieval Warm Period of hundreds of years ago in which (they say) temperatures averaged much higher than during most of the surrounding centuries. They argue that data from the

Little Ice Age should also be added, because such information clarifies the considerable ups and downs of global temperature, sometimes lasting for hundreds of years, whether or not humans are significantly interfering with the atmosphere.

They assert that when the Medieval Warm Period is incorporated into the graph, it demonstrates that there were no disastrous consequences such as massive floods, droughts, habitat destruction, spikes in severe weather, and loss of coastal lands flowing from a significant and protracted temperature increase. Further, this long period of higher mean temperatures preceded the Industrial Revolution and all modern forms of carbon-emitting technology by hundreds of years. They published their own less-frightening version of the hockey stick graph, which they offer as evidence that recent temperature trends, while genuine at some level, are nothing we haven't experienced before, and that humans may not even be the source of the current modest increases.

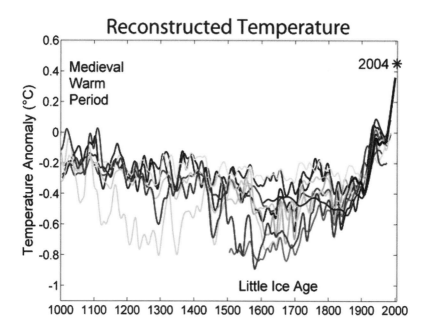

You don't need to be a rocket scientist, a hockey player, or even an old entomologist like me, to understand that it matters whether the hockey stick graph is generally valid. If it is, then we are now riding a meteoric surge in world temperature the likes of which haven't been seen during recorded human history. In that event, we could be lighting the fire that will destroy us, burning down our house with ourselves locked inside it. Never a good idea. But if the hockey stick graph isn't an accurate view of climate change patterns, then the world has, fairly recently, literally weathered periodic natural fluctuations in temperature at least as large as our current trend and for hundreds of years at a time, and it did not spell serious trouble for our ancestors. That's a huge and impactful difference in meaning that hinges on whether we accept the hockey stick graph or an irregular alternative. This battle of the graphs is one of several instances in which the climate change debate has seen people divide into opposing camps that can't even agree on some basic facts.

When highly educated and experienced experts refer to themselves by such sports terms as "the hockey team," it's a signal that the debate has moved away from the purely objective, scientific realm. It's an early warning sign that participants in the controversy themselves view it as a type of game or athletic competition rather than a fact-based, principled inquiry. Ever since the hockey stick graph was first published, "players" on both sides of the red line have been clashing with one another, wielding their sticks like cavemen's clubs and not like professors' pointers. And the hockey game has featured congressional inquiries, United Nations analysis, National Academies of Sciences review, and plenty of collateral damage.[5] We will return to this very long, very unruly hockey game a little later in this chapter, but first we need some perspective on possible causes and effects of climate change. In other words, if it's happening, did we do it, and is it anything to work up a sweat about?

Assuming that there is a modern net increase in atmospheric hydrocarbons, brought on by our modern industrialized activities that depend so heavily on combustion of fossil fuels like coal and oil, what does it have to do with climate change? This is a complex and elusive question, to say the least. It has been the subject of numerous entire books, on both sides of the divide as to whether we are causing harmful climate change, and I don't want to get into the tall weeds on the evidence for and against. It will suffice to give a brief taste of the bewildering maze of competing paths an investigator might confront as he or she attempts to understand our current climatic situation.

To give just one example of this complexity, the Earth's climate is dependent on an intricate and incompletely understood tangle of variables. It would be overly simplistic to presume that an increase in atmospheric carbon dioxide will necessarily produce the same conditions in the world's laboratory that it might provoke in a scientist's laboratory under carefully controlled conditions. In the real world, climate is influenced not only by the total amount of greenhouse gasses in the atmosphere, but also by such factors as the concentration and location of those gasses, the number and type of clouds (condensed water vapor) generally present, the overall level of humidity, and the possible adaptation of Earth's vegetation in response to higher levels of carbon dioxide.

These and other factors are not independent of one another, but rather a change in one can exert influence on others that might tend to reinforce or oppose the effects of any single factor. There are feedback mechanisms by which changes in some factors can produce responsive and often countervailing changes in others. If global carbon dioxide is increasing, for instance, this might spur greater than normal growth and reproduction of photosynthetic plants, which in turn could remove more than the usual quantity of carbon dioxide from the air and convert it to oxygen. The net impact on mean global temperature under such circumstances is not easy to predict.

It's very important to recognize that global temperatures always represent a rough, not terribly precise average, an approximation of the overall world temperature. This single number has to do a lot of heavy lifting, and it often shows the strain. It includes within its one figure all the variation from region to region, from pole to pole, from land to sea, from desert to jungle, and from altitude to altitude. Temperature can be rising significantly in some areas (as it has been doing recently in the Arctic regions, for example), while remaining stagnant or even declining in others (as in much of North America, the western Pacific, and the Arabian Peninsula).

Global mean temperatures are calculated in part using actual readings from a large but limited number of measuring points. As of this writing, the global temperature-monitoring network included 517 weather stations, and from these 517 outposts samples are collected at various intervals. Scientists and technicians then use computers to extrapolate from the available sampling numbers and fill in all the gaps to arrive at one number that will represent the whole world's temperature.[6] These sampling stations are scattered around, and not too evenly, with more in some areas and far fewer in others. Very remote, very frigid, and very dangerous areas (such as the frostbite inducing places near both poles) typically receive less coverage than nicer ones. Go figure. The regional variation in temperatures as measured by these stations (and by satellites now) goes along with significant fluctuations from month to month, year to year, and decade to decade, even when there is a long-term global trend that points up, down, or flat. Suffice it to say that taking the temperature of a patient as large and hefty as our planet entails more than a few challenges.

Given such complexity, it will be sufficient for purposes of this book to cover some of the basic aspects of the evidence in very abbreviated, summary form. I don't claim that this treatment is in any sense comprehensive—only that it will serve to illustrate the gaps and uncertainties in the scientific and factual record that have led to prolonged legal inertia worldwide. If you don't want to step into the deep mud, you can safely skip the next several pages, because all I want to show is that there is some credible scientific evidence and doubt on both sides of the climate change hockey game. That's really all I need to establish to set the stage for my eventual introduction for a system that allows us to rationally work through the risks, rewards, and probabilities. So if you're still with me, and with this preface in mind, here are a few data points to consider.

According to National Aeronautics and Space Administration (NASA) figures, on balance the Earth's surface temperature has risen nearly 1.5 degrees Fahrenheit or 0.75 degrees Celsius during roughly the last 100 years. Based on one view of the available data, the trend has been accelerating, with 11 out of the 12 years between 1995 and 2006 on the list of the warmest years since reliably accurate recordkeeping began in approximately 1850. Much of the raw information comes from very incomplete samples unevenly distributed around the world, at least during the first half of this recent period, so there is some doubt as to the validity of such definitive conclusions. Nonetheless, during more recent years the available samples of Earth's surface temperature have been considerably more representative of the planet as a whole, so there are much stronger indicia of reliability and comprehensiveness for these readings than we have for the earlier estimates. The chart below depicts NASA figures on the trend for the last 100-plus years, for which we have better and more direct data than the more distant "cold case" centuries.[7] Are there signs that these elevated recent temperatures are having any detectable effects?

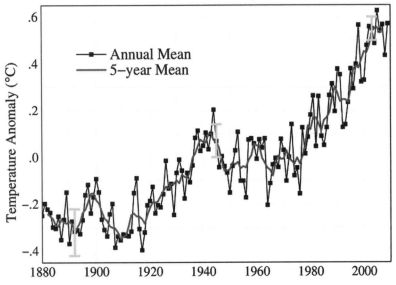

Global Land–Ocean Temperature Index

Some of the immense ice sheets in both Greenland and Antarctica have receded significantly during recent years in both area and mass, perhaps due to melting caused by rising ambient temperatures near the poles. This Arctic and near-Arctic regional temperature increase appears to have continued at a notable level even during the last 10 years or so, when much of the planet may have enjoyed flat or slightly declining temperatures.[8] There is also some evidence that mountain-top glaciers and high-altitude snow mass may have been shrinking in other parts of the world. Evidence gathered by the NASA Jet Propulsion Laboratory's Gravity Recovery and Climate Experiment indicates that Greenland lost between 36 and 60 cubic miles (between 150 and 250 cubic kilometers) of ice per year between 2002 and 2006, and that Antarctica shed about 36 cubic miles (152 cubic kilometers) of ice between 2002 and 2005. When that much ice melts, there's arguably a lot more liquid water around that has to go somewhere, and that somewhere is the world's oceans. NASA data demonstrate that sea levels worldwide have risen by about 6.7 inches (about 17 centimeters) during the last century, with the pace of rising waters nearly doubling in the last decade.

Up to now, the rise in sea level has been small enough that most people are entirely unaware of it. Like average temperature itself, sea level has been rising and falling for millions of years, long before humans arrived, and so there is again the question of whether recent changes are merely part of this grand natural cycle or something human-induced and different in both kind and degree. But if the recent trend toward climbing sea levels continues, or accelerates, and if there are no natural countervailing forces that act to lessen or reverse this phenomenon, there could be harmful consequences far beyond the obvious flooding of low-lying coastal areas. We could see a meltdown truly worthy of the name. For example, the influx of salt water from climbing ocean levels could decimate the supply of potable water in many near-coastal centers of human habitation. There is evidence that salt water is already contaminating some of the world's most productive agricultural areas, such as China's Yangtze Delta and Vietnam's Mekong Delta, as well as underground water sources in Israel, Thailand, and various small islands throughout the world.

Some scientists also have observed recent trends towards more extreme and powerful storms and hotter, longer dry periods, with a possible link to global climate change. This is because higher temperatures can cause increased evaporation, and a warmer atmosphere can retain more moisture, resulting in more water in the air that can fall somewhere as precipitation. The frequency of heavy-precipitation events has gone up in most land areas. There is evidence of significantly increased precipitation in eastern parts of North and South America, northern Europe, and northern and central Asia, plus an increase of intense tropical cyclone activity in the North Atlantic since about 1970.

However, under this view of the evidence, regions that are already arid will probably lose even more water if the weather is hotter. This causes worsening of water shortages in dry areas, through accelerated and intensified droughts and desertification, i.e. land turning into desert. Scientists have recorded increased drying in large areas such as the Mediterranean, southern Africa and parts of southern Asia. Notably, in Africa's large basins of Niger, Lake Chad, and Senegal, total available water in recent decades has decreased by 40 to 60 percent. Likewise, the inexorable spread of desertification in arid regions has been spurred on by declines in average annual rainfall, runoff, and soil moisture in many parts of Africa.

In this sense, climate change might add to the problems of both severe storms and drought, in a climate-oriented version of the cliché "The rich get richer and the poor get poorer." It also, however, provides ammunition for use by climate change's skeptics, who can point out the apparent contradictions inherent in a theory that blames rising greenhouse gas emissions for simultaneously causing too much precipitation and worsening droughts. Along these lines, I must point out that there is an exquisitely nuanced interplay between surface warming and changes in cloud cover, and the *possibility* that clouds generally might display temperature-induced modifications that serve to counteract global increases in surface temperature. Such types of *negative feedback effects* might help to keep the planet's climate more or less stable despite perturbations in some environmental conditions, at least under some circumstances. On the other hand, it's also possible that *positive feedback effects* might magnify and amplify the harmful impacts from certain environmental changes.

Anecdotally, there are indications that some species are migrating toward either shrinking colder regions or toward expanding warmer habitats in a noteworthy departure from their long-established ranges. Species that were previously unknown in some regions have begun appearing there, while long-familiar living things have vanished from their old stomping (or crawling, or slithering) grounds. If climate change is driving such migrations of insects, mammals, birds, reptiles, and other life forms, there could be profound ripple effects. Ecosystems could gain or lose key components, and food webs might be stripped of key supporting players. In a similar manner, there is some evidence that climate change is producing longer growing seasons, earlier seasonal mating and egg-laying, shortened hibernation, shifts in seasonal migration patterns, and other behavioral and physiological transformations. The long-term ramifications of these types of large-scale biodiversity modification are very difficult to predict, and may be extreme.

Overall, at this point in time the effects (if any) of climate change have been gradual, and indeed virtually imperceptible from the ordinary human perspective. The best available measurements do seem to indicate a small but real and measurable rise in average temperatures (with plenty of short-term periods of flat or declining temperatures mixed in) in a wide array of locations during the past two centuries. However, the planet has experienced rather extensive episodes of significant increases and decreases in its mean temperature before, many times, and even long before people were around or burning large quantities of fossil fuels.[9] During roughly the past 650,000 years, seven major instances have occurred during which glaciers generally grew and spread outward from both poles, and then eventually retreated again. There have been ice ages and periods of much greater warmth, lasting for hundreds of years, in a grand cycle of fluctuation, with the most recent full-scale ice age ending approximately 7,000 years ago. There have been significant variations in atmospheric concentrations of greenhouse gasses as well, accompanying these very long-term temperature trends.[10] However, according to some sources, the increases in atmospheric carbon dioxide and methane since about the year 1750 amount to percent rises of 36 and 148, respectively, and have resulted in concentrations unprecedented during the time span we can reliably assess.[11]

Given these natural historic ebbs and flows of global mean temperature, where periods of both warm and cold "anomalies" can persist for centuries, we need to be careful about making too much of any short-term patterns. These temperature anomalies are divergences up or down from some point of reference we select as the norm. We know some anomalies have lasted for hundreds of years, and did so long before people began adding appreciably to the greenhouse gasses in the atmosphere. So even a few consecutive decades of more or less steady movement either up or down do not necessarily reflect anything out of the ordinary, or any anthropogenic meddling with the planet's thermostat. The next graph illustrates a comparison of temperature trends during the three most recent decades, as measured or interpreted by three different well-known sources of climate information.[12]

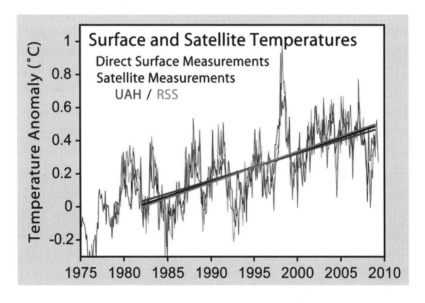

The temperature trends from these three very modern sources all point in roughly the same climbing direction, but with variations that might be surprisingly large to some observers. The slope of the line varies from 0.163°C/decade for the University of Alabama at Huntsville, to 0.187°C/decade for the surface readings, and 0.239°C/decade for the Remote Sensing System. Which would you consider more significant: that all three sources show an upward trend, or that the divergences could amount to a difference of several degrees if continued for a century or more?

Is the current trend worrisome? Does it deserve to provoke a meltdown reaction from a united human race? In the past, these cycles of glacier advance-and-retreat were very likely attributable to natural phenomena unrelated to human activity, such as slight fluctuations in our planet's orbit around the Sun, which in turn either increased or reduced the amount of sunlight/solar energy reaching us. Are we the primary cause of this contemporary warming trend, or is this just one more in a long series of natural ebbs and flows in Earth's temperature? Will the subtle but measurable warming trend of the past century or so continue, or accelerate, absent meaningful, or even drastic, human countermeasures? If so, will there be significant melting of the polar ice caps, disruption of vital ocean currents, dangerous elevation of sea levels in heavily populated areas, adverse modification of many key habitats, massive crop failures, elevated incidence of hurricanes, droughts, and other extreme forms of harmful weather, or additional grave impacts? If so, are such catastrophes imminent, near-term, intermediate-term, or hundreds of years off in the future?[13]

One reason why we've taken so little legal action to curtail any looming hazards from global warming is that many policy makers subscribe to the shopkeeper's philosophy, "You break it, you bought it." They also adhere to the converse principle: "If you didn't break it, no one can force you to pay for it." The blame-fueled view of legal responsibility places heavy emphasis on causation as a prerequisite to taking legal corrective action. It holds that human beings should not be made legally bound under environmental law to make changes in the way they do business, conduct their daily affairs, or generally lead their lives unless there is substantial proof that people cause the problem in the first place. In this view, only anthropogenic climate change (the kind we bring upon ourselves) is even arguably deserving of our intervention.

I have personally encountered a close relative of this perspective many times within my own household. If I ask one of my two children—either Christie or Julie-Kate, take your pick—to pick up some large heap of tangled, mangled, and disorganized toys that is on long-term display decorating the middle of our living room floor, there is a high probability that one or both of the girls will reply, with an air of satisfied finality, "I didn't do it! You didn't see me do it!" In their agile and creative minds, their self-assessed finding of insufficient proof of a causal link definitively negates any possible duty to do anything to clean up a mess or, depending on one's perspective, a valuable example of modern art. There are at least two very important corollaries that follow from this axiom: (1) Messes invariably cause themselves; and (2) My daughters should never have to clean up any mess, whatever its origin.

Presumably, this analytical framework would remain intact no matter how immense the haystack of randomly-juxtaposed smirking Bratz dolls, Barbies, Lego/Duplo bricks, and assorted stuffed plush mammals grew, and/or how many years it might continue to encroach upon our family's central supposedly living space. Many times I have made a brilliantly rational, logical, and rather passionately delivered argument that this ever-expanding toy mosh pit poses a serious tripping hazard to everyone in our family, including both kids and our Abyssinian cat, Abby Sunstein, in an attempt to appeal to the girls' sense of moral duty. This exercise reliably ends with me in a state of enhanced exhaustion and elevated frustration, and the toys in a gloriously undiminished pile of prolific profusion. It is evident that Christie and Julie-Kate have serendipitously derived a fundamental tenet of international and domestic law, because very much the same set of principles has been used to argue effectively against any meaningful legal response to climate change's dangers. If no one can prove conclusively that we did it, why should we inconvenience ourselves to do anything to undo it? So what if the condition seems likely to harm us and millions of others? The situation lends itself well to a small bit of doggerel: If we have no blame, we'll let things stay the same, even if the hazard takes aim at our name.

Whether humans are primarily to blame for any significant part of the contemporary climate change situation is actually irrelevant, on an objective level, if the other factors point toward trouble. In my view, a person should "bet" in accordance with the probabilities associated with any particular situation, regardless of who's responsible for bringing that situation about. A predicament, or even a deadly dilemma, is no less serious a threat when caused by forces beyond our control; what should matter in every instance is whether we can and should do anything to improve our plight. Irrespective of the precise dimensions and nature of our role as one of the causal factors provoking climate change, it would still be incumbent on people to mobilize whatever resources we can to ameliorate the problem.

Is it in our self-interest to avert potentially disastrous climatic shifts, even if we had no responsibility for bringing that change about, for the simple reason that we have so much to lose? Moral responsibility aside, we may have good reason to take action *if* we are seriously threatened by this or any other major environmental phenomenon, and *if* by our rational calculation the costs of taking action are outweighed by the magnitude and likelihood of the harm that would happen if we do nothing. This seemingly obvious point is often obscured in the confusing and contentious debate persistently raging, hockey sticks swinging high, around so many aspects of the climate change controversy.

I will return to this key argument later in the book, because uncertainty about causation stands alongside uncertainty about danger as a major and persistent obstacle to effective corrective action on the greatest environmental threats ever to face this planet. It is enormously important that we make our biggest decisions with rational and logical assessment of what we stand to gain and what we might lose, depending on the course of action we select. It is also crucial that we make proper use of the odds, the probability that various good or bad events will follow from the decisions we make, and the probability that the outcomes will be very large or very small. More on this will come later.

Legal attempts to confront global climate change have been sporadic at best, but there has been no shortage of superheated rhetoric and publicity-hunting on both sides of the divide. Torrents of talk aside, the United States has not been a leader in meeting the climate change challenge, in part because of skepticism as to the reality and magnitude of the threat, and in part because of fears that American industries and citizens would bear an unfair and disproportionate share of the burden while other large nations such as China and India remain free to continue business as usual and gain a major competitive advantage.[14] American political leaders have been able to refuse to participate in such initiatives as the Kyoto Protocol, largely because they could rely on the potent combination of self-interest and scientific uncertainty.[15]

But there's more than one flavor of uncertainty, and more than one avenue toward legal action on the environment! In case you missed it, the United States Supreme Court handed down one of the most far-reaching decisions in history in April 2007.[16] In the case of *Massachusetts v. Environmental Protection Agency*, the nation's highest court held that EPA was authorized to regulate greenhouse gasses as a "regulated pollutant" under the Clean Air Act.[17] The Court ruled that, in the event EPA issued a formal determination that greenhouse gasses cause or contribute to climate change that endangers the public health or welfare, EPA had the authority by virtue of the Clean Air Act to promulgate appropriate regulations for new motor vehicles in the United States.

After surveying the evidence and receiving some 380,000 public comments, EPA issued its "endangerment determination" on 7 Dec. 2009, as well as the related "cause or contribute" determination.[18] Together, these two findings constituted EPA's official legal conclusion that the *combined* emissions of six greenhouse gasses from our new motor vehicles and their engines contribute to "greenhouse gas pollution" in the atmosphere which threatens the public health and welfare of current and future generations.[19] The specific greenhouse gasses covered by these findings were carbon dioxide, methane, nitrous oxide, hydrofluorocarbons, perfluorocarbons, and sulfur hexafluoride (the first three of which can be naturally occurring).

By the terms of the Clean Air Act, the EPA's findings make it possible for EPA to regulate emissions and/or atmospheric concentrations of these six gasses without any legislative action from Congress or any direction from the White House. In fact, EPA may be legally *required* to do so, and might well be compelled by citizens' suits to issue these regulations even if the Administrator doesn't want to. And although *Massachusetts v. EPA* dealt specifically only with motor vehicle emissions, once greenhouse gasses become a "regulated pollutant" the statute compels regulatory action for many stationary sources as well.

I won't go into much detail, but the fact that the six listed greenhouse gasses (in any combination or alone) are now officially a "regulated pollutant" under the Clean Air Act could easily be interpreted to *mandate* (and not merely to allow) several very significant legal actions by EPA. For example, Section 111 of the Clean Air Act requires EPA to regulate certain so-called "major" stationary sources of any "regulated pollutant," with "major" defined by the statute as the potential to emit more than 250 tons per year of such a pollutant. New or significantly modified "major" stationary sources in certain areas require "new source performance standards" and preconstruction review, using "best available technology" to reduce emissions. Also, Title V of the Act calls for operating permits and certain controls on emissions for sources with the potential to emit that same 250 tons per year threshold. Many thousands of sources of greenhouse gasses that have never been regulated under the Clean Air Act before will far exceed this threshold, and could now require legal restrictions.

Plus, new motor vehicles or new engines must have emissions standards for "regulated pollutants," under Section 202 in Title II of the Clean Air Act, and now could need technological controls for their greenhouse gas emissions. Arguably, EPA is even required under Title I to issue National Ambient Air Quality Standards (NAAQS) for "regulated pollutants" of this type. If EPA promulgates NAAQS for these six greenhouse gasses, states will have to follow suit with provisions in their State Implementation Plans to monitor and regulate activities so as to bring Air Quality Control Regions within their borders into compliance. Failure to come into attainment with the NAAQS triggers legal consequences, regardless of whether non-attainment is primarily the result of the pollutant blowing in from other states or nations, or even comes mostly from natural sources.

As of this writing, there were some signs that EPA had realized the awesome magnitude of its new regulatory power brought about by the *Massachusetts v. EPA* decision. Possibly out of fear that aggressive regulatory action within the usual interpretation of the Clean Air Act would be politically damaging, economically crushing, or at least extremely hard to manage as a practical matter, EPA was proposing a "tailoring rule." This novel device would let the administrative agency re-define, without any act of Congress, the threshold of greenhouse gas emissions that triggers the various provisions of the Clean Air Act, changing it from the usual 250 tons per year to 25,000. Again, EPA may not have this unilateral legal power, and a citizen's suit could force it to apply all the standards under the normal 250 tons per year threshold. Time will tell, as litigation will hammer out solutions to this problem and the other uncharted contours of the hazy new legal situation, one case and one issue at a time.

All of this legal uncertainty springs from our lack of a well-planned, organized, politically accountable approach to climate change. In the legal vacuum left behind by congressional inaction, courts and administrative regulatory agencies have stepped in. No one has ever confused me with any of the prophets of old, but I think I can safely predict that we have many years of litigation ahead of us before we can know the full meaning of *Massachusetts v. EPA*.

Many scientific studies, including those summarized here and others that formed the basis of the holding in *Massachusetts v. EPA*, have credibly concluded that climate change is a real contemporary phenomenon, with reason to be concerned about its ramifications. There are varying views, but a sizable body of work supports the reality of global warming as a matter well worth our attention, even in the absence of absolute (and likely unattainable) scientific consensus.[20] While allowing for years and even decades during which global mean temperatures may remain roughly constant or even slip downward a bit, a substantial body of data supports a total planet-wide increase of about 1.25 degrees Fahrenheit (0.7 degrees Celsius) from the 1970s to the late 1990s. Depending on our choice of evidence, this trend apparently then leveled off considerably, to an overall increase of between just 0.2 and 0.07 degrees Celsius from 1999 to 2008.[21] After factoring in such periods of ups and downs, spikes and stagnations, and all sorts of regional and short-term variations, a credible case can be made that the planet as a whole has experienced a measurable and *possibly* noteworthy climb in temperatures for the most recent century or more.

On the other hand, although less attention has generally been paid to them, major works by learned and well-credentialed skeptics also maintain that any recent fluctuations in the planet's temperature are well within normal limits from a long-term historical perspective.[22] This creates a complex problem. There are some significant doubts and uncertainties in the underlying scientific and factual foundations upon which climate change theory is based. We will now briefly summarize a few of these to get the issues out in the open. Then we'll be prepared to deal with them appropriately.

The notion that climate change poses a serious threat to the future of the planet has been, arguably, seriously exaggerated. The crux of the climate change debate surrounds the idea that it is anthropogenic, and can only be halted by massive legislative initiatives in a form typified by the American Clean Energy and Security Act, the bill known as Waxman-Markey that passed the House of Representatives and then languished in the Senate. This and similar proposals, if enacted, could create substantial consequences for the U.S, economy and the liberties of American citizens. Skeptics argue that climate change theory relies on data inconsistent with real-world factors, and that it calls for large-scale changes allowing the government a heavy hand in regulating society unnecessarily.

For centuries the Earth has experienced intermittent periods of temperature change consistent with natural factors, including ocean currents, volcanic eruptions, changes in the Earth's orbit, and the amount of energy released from the Sun.[23] On a rudimentary level, the science behind climate change involves the natural phenomenon in which carbon-based fuels, when burned, give off greenhouse gases that create a warming effect in the Earth's atmosphere.[24] These greenhouse gases are critical for the survival of life on Earth, retaining the Sun's radiant energy that would otherwise escape into space, uninterrupted.[25] Carbon dioxide accounts for just over a quarter of the atmosphere's composition, only one ten-thousandth more than it did 250 years ago.[26]

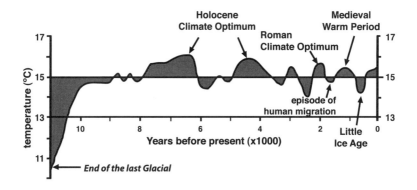

The graph above depicts at an extremely general level of detail the average surface temperatures of the northern hemisphere during the past 11,000 years. Until recently, the variations in temperature, including the Medieval Warm Period and other relatively hot times, could not have been caused by anthropogenic carbon dioxide emissions, given the time period and the lack of industry. If these figures are reliable, the evidence tends to verify the notion that additional (non-human) factors can be responsible for changes in the climate.[27]

The United Nations Intergovernmental Panel on Climate Change (IPCC) is the leading international environmental organization that advocates worldwide implementation of legislation to combat climate change. The IPCC bases its predictions about the changing state of the Earth's climate in part upon computer models that estimate the rate of global warming. Some critics, however, argue that the IPCC models are imperfect due to the exaggerated estimates used as variables in achieving these predictions.[28] A 2009 report based on peer-reviewed scientific studies illustrated the possible disconnect between the actual propensities of the climate and the artificial IPCC computer models that assume fundamental data when attempting to predict the greenhouse effect.

For example, water vapor is a dominant greenhouse gas. It's controlled by the laws of physics such that a rise in temperature will increase the amount of water vapor that the atmosphere can hold.[29] The IPCC models all work on the assumption that global relative humidity is constant, and do not account for any slight deviation consistent with the ever-changing variable of water vapor in the atmosphere, which could alter the expected response of an increase in carbon dioxide. Some data suggest that relative humidity has, in fact, fallen in the past 60 years, which could indicate a decrease in the atmosphere's carbon dioxide levels.[30]

Assuming, for sake of argument, that the IPCC models reliably calculate the rate of global warming, they could nevertheless fail to accurately predict climate change, according to some. This is because the models do not take into account the climate's chaotic nature, a well-known characteristic acknowledged even by the IPCC itself. Edward Lorenz, in his momentous chaos theory, argued that long-run climate prediction is impossible absent precise knowledge of the initial state of myriad variables that define the climate. According to chaos theory, the climate is so complex and sensitive that any deviation from the correct initial variables, however minor, would disrupt the whole system entirely, causing a completely different outcome. Without *exact* information about each variable identified with complete accuracy, any definitive prediction as to the climate's future is simply not feasible due to the sudden changes of state of the variables that could intervene at any time and in any fashion.[31] With this in mind, we'll look at chaos theory in more detail in Chapter 7.

Improvements in geospatial technology have also called into question some aspects of climate change theory. Global Positioning System equipment and data processing have allowed scientists to closely observe ice-sheet movement by positioning stakes in the ice sheets. Recent measurements made along the "K-transect" of the Greenland Ice Sheet, just north of the Arctic Circle, have apparently shown a significant slowing of movement in the past 17 years, resulting in a 10 percent decrease in the annual average velocity.[32]

The National Snow and Ice Data Center reported findings that Arctic Sea ice had expanded by 13.2 percent in 2007 alone.[33] To illustrate this growth, 13.2 percent of the Arctic Sea ice translates into an area just larger than the State of Texas, or roughly 270,000 square miles.[34] Climate change theorists have addressed these data to harmonize the evidence with their model.[35] However, some skeptics assert that it defies logic to conclude that both an increase in temperatures, and a decrease in temperatures, are indicative of "global warming," and that it is disingenuous to respond to this inconvenient point by renaming the phenomenon as "climate change."

There may also be a problem arising from the lack of a fully free and open discussion of climate change on the merits. There are some reports indicating that numerous scientists have expressed their doubts as to the climate change phenomenon, but have been met with opposition. According to Jonathan Manthorpe of *The Vancouver Sun*, many "scientific and academic voices have fallen silent in the face of environmental Jacobinism."[36] For example, when climatologist Dr. Timothy Ball criticized global warming theory in a documentary entitled "The Great Global Warming Swindle," he received hate mail. One message in particular read, "If you continue to speak out, you won't live to see further global warming."[37] Similarly, Ian Plimer, geologist and author of *Heaven and Earth—Global Warming: The Missing Science*, a best-selling, prize winning book that challenged global warming theory, had difficulty finding a publisher for his more recent work. He believes this was attributable to what he sees as the intimidation tactics of the environmental lobby.[38]

There are other reports of unfree speech in this area. When Dr. Alan Carlin of the Environmental Protection Agency (EPA) prepared a spring 2009 report that questioned the need for regulation of carbon dioxide and expressed concern as to the scientific reasoning behind anthropogenic global warming, the EPA wouldn't release it.[39] Dr. Carlin asserts that the EPA banned him from writing or speaking about climate change, and demoted him to performing data entry for the agency. One of the many who spoke out against the "settled science" of climate change proponents, Dr. Carlin stated that he felt as though he "was under a gag order."[40]

Critics argue that proponents of climate change who discourage investigation and silence opposition are not searching for scientific truth, but setting forth a dogma. They lament the loss that they believe science will sustain if some groups continue to stifle dissent. They maintain that science can only flourish when truth is paramount and propaganda is ignored.

Some climate change theorists have argued that there is no need to investigate their thesis's validity because a settled consensus exists among the scientific community reflecting overwhelming support. This idea, however, is arguably inconsistent with some of the available evidence. One survey of 400 of the 1,000 individuals worldwide who have made climate study their profession found that 90 percent were not convinced that we are witnessing anthropogenic climate change. A 1997 survey of American climatologists from 36 states showed that nine out of 10 agreed that "[s]cientific evidence indicat[ing] variations in global temperature are likely to be naturally occurring and cyclical over very long periods of time."[41] The Manhattan Declaration of 2008 brought together hundreds of prominent scientists who disavowed the so-called "consensus" among the scientific community that climate change proponents assert. The declaration stated, "there is no convincing evidence that carbon dioxide emissions from modern industrial activity has in the past, is now, or will in the future cause catastrophic climate change, [and] . . . attempts by governments to inflict taxes and costly regulations on industry and individual citizens with the aim of reducing emissions of carbon dioxide will pointlessly curtail the prosperity of the West and progress of developing nations without affecting climate."[42]

Much depends on who qualifies as a "scientist" or an "expert" in climatology when we're counting heads (even eggheads) and taking opinion polls in search of scientific consensus or lack thereof. If we have a particular point of view, and we're eager to see that view accepted as settled consensus, we can try to define our opponents out of the picture. By our own self-selected definition, then, their views won't count, and won't disturb the consensus we're hoping to forge.

Let's look a little closer at some of the recent temperature figures for a dramatic illustration of how data from the same period has been interpreted in—if you'll pardon the pun—polar opposite ways. Those who oppose the idea of dangerous warming trends argue that global temperatures have actually declined in the past 10 years (at least in some large parts of the world), and that, since the 1970s, North America has experienced record cold temperatures.[43] If reliable, such data could (very recently) demonstrate the movement of Earth's average temperature in a downward direction and illustrate the trend of what *might* be interpreted as global cooling. Or it might be a brief respite from a prolonged and perilous upswing in global temperatures. Who's got the hockey puck?

The period of time at issue here is just a few years, and it is normal for the global temperature to bounce up or down a bit for such relatively short periods, even during much longer, protracted trends of increasing or decreasing temperatures. It is objectively impossible to draw definitive long-range conclusions based on readings gathered over the span of a few years, or even a decade or two, when one considers the vast spans of time over which Earth's climate fluctuates naturally. Still, it's worth noting the data, and being aware of possible varying interpretations (as usual).

On the opposite side of the global warming dispute, the IPCC has projected a significant increase in global temperature using meteorological station measurements of sea surface temperatures, coupled with an array of satellite measurements. The IPCC stated that it was

90 percent confident that emissions of carbon dioxide by humans were the cause of global warming.[44] The IPCC's Fourth Assessment Report, discusses data from the same period as the above studies, and instead of a cooling trend projects a future sharp *increase* in global temperatures, consistent with the anthropogenic climate change hypothesis.[45] In light of the questions raised initially in late 2009 regarding the data reliability from the Climatic Research Unit (a key source of evidence for the IPCC), some would suggest an extra dose of scrutiny is appropriate here. Similar trends, however, have been identified by other major governmental entities, including NASA, NOAA, and EPA, so it isn't appropriate to jump to instant and sweeping conclusions.[46]

EPA scientist Dr. Alan Carlin believes that one reason for the divergence of the IPCC projection and other scientists' contrary opinions is that the IPCC models do not take into account solar variability. According to Dr. Carlin, changes in the sun's eruptional activity, solar wind, and magnetic field are some of the major factors affecting global temperatures, and have not been incorporated in the IPCC prediction models. Further, Dr. Carlin states that the IPCC models are inadequately equipped to predict major global climate patterns due to the insurmountable number of variables contributing to the climate's complex nature (in addition to solar variability), and the models are only as good as the data upon which they rely.[47] This is not to say that the argument supposing global climate change is in error, but simply implies that more factors may cause climate change than just human-generated carbon dioxide and other greenhouse gas emissions.

Dr. Carlin takes into consideration that much of the debate climate change's cause arises from the fact that we must wait until future temperature readings can be made to determine definitively whether we will actually be thrust into a period of either global warming or cooling. If global temperatures decrease despite increasing greenhouse gas emissions, we could assume that global warming advocates may need to alter their hypothesis. If, on the other hand, temperatures increase while solar activity declines, skeptics of anthropogenic global warming would have to adjust their argument.[48] Because such crucial uncertainties abound, it is valuable for us to explore the possible economic effects and societal changes that would be experienced for each scenario. In an unclear situation, the prudent course of action may differ greatly depending on whether any mitigation measures are easy and affordable or onerous and exorbitant.

The legislation suggested by environmental advocacy groups, as well as regulations promulgated by EPA in the aftermath of *Massachusetts v. EPA*, aim to curtail carbon dioxide and other greenhouse gas emissions as a means of proactively arresting potentially disastrous climate change. But some skeptics claim that combustion of fossil fuels and other greenhouse gas-generating processes are crucial components of a thriving economy, and that their restriction will have a crippling effect on the nation's productivity and quality of life.[49] According to William Nordhaus, author of *A Question of Balance: Weighing the Options on Global Warming Policies*, in order to achieve "optimal" limitations on temperature increases to 1.5 times their pre-industrial levels, the United States would have to spend over $27 trillion. Nordhaus assumes that this is technologically feasible and focuses only on the cost—over twice the national debt, or greater than $38,000 per United States citizen.[50] Former Vice President Al Gore's more aggressive proposal that emissions be reduced by 90 percent by 2050 would result in an expenditure of $33.9 trillion by the United States.[51] The implementation of such vast reforms would impose a severe burden on the United States economy that would extend to taxpayers and businesses alike. Critics of climate change intervention maintain that such profound economic sacrifice is a startling and unthinkable notion, especially if we consider the evidence arguably calling into question anthropogenic climate change.

In terms of pollution in the global arena, the United States is by some measures not the "bad guy." In fact, it can be argued that quite the contrary is closer to the truth. Here's one way to put it in perspective. When looking at a comparison of the ratio of economic production to greenhouse gas emissions, the United States is the most efficient. It produces $2,000 in economic value per ton of greenhouse gas, compared to Indonesia's production of $679, India's $497, and China's $450.[52] Of course, other metrics would point to a different conclusion. Nevertheless, skeptics question both the effectiveness and fairness of strict reforms to be implemented in the United States, curbing its economic activity, while some other major sources of emissions (such as China, Brazil, and India) are able to continue producing both goods and gasses, and remain largely unregulated.[53] If the United States were to submit to some of the more ambitious climate change reforms, it would arguably run the risk of jeopardizing its premier economic status and the well being of its people.

In light of this, is there enough respectable, if not actually respected, authority on both sides of the climate change divide that we can say this is, in truth, an issue with no obviously clear-cut right and wrong resolution at present? On the side of a serious climatic emergency, an accumulation of credible evidence exists that proponents claim should spark reasonably prudent leaders to develop and implement effective political and legal action to meet the potential threat.[54] Some notable and well-respected scientific organizations have issued public statements in an effort to move the issue into the action phase.[55] The United Nations Framework Convention on Climate Change was an attempt to launch a comprehensive international response, and it attracted enough signatories to enter into force.[56]

But skeptics within the United States and some other key nations have focused on the causation questions, inaccuracies in the data, alternative interpretations of the evidence, large historical temperature fluctuations, and allegations of environmentalist alarmism. To the detriment of a reasoned and reasonable resolution, some prominent climate change activists and skeptics alike have alleged ill motives, falsification of evidence, and willful denial of reality on the part of their opponents. Thus the debate has roared on, replete with cartoon-like exaggerations and *ad hominem* attacks all around while little or nothing of practical effect was accomplished.[57]

The climate change controversy sometimes looks a lot like a war. There is espionage, demonization of opponents as enemies, disinformation, destruction or concealment of evidence, suppression of enemy communications, and a ruthless win-at-all costs attitude—at least on the part of some "combatants"— more appropriate for armed conflict than professional scientific inquiry.

When computer hackers illegally uncovered hundreds of e-mail messages and other documents of Britain's Climatic Research Unit in late 2009, some of the texts provided damning evidence of the extremes to which certain prominent partisans have been willing to go. Some e-mails pointed to politically motivated evidence selection and manipulation, concealment and even destruction of underlying data, deliberate doctoring and twisting of evidence, suppression of opposing viewpoints, distortion of the peer-review process to exclude contrary authors and their work from publication in respectable journals, and attempts to punish and marginalize editors and journals willing to publish skeptical findings.[58]

The impression created by the documents was that of an organized campaign by some very highly-placed and influential individuals to steer the climate change science solely in the direction of an alarmist conclusion. With the Climatic Research Unit (CRU) at the University of East Anglia officially serving as one of the most important major sources of information for the U.N. Intergovernmental Panel on Climate Change, allegations of such widespread deception, suppression, and intentional manipulation within the CRU struck like a lightning bolt.[59] If some of the most powerful climate-change scientists were involved in a massive perversion of the scientific method to suit their own agenda, what did this mean for the entire debate? The mess was quickly dubbed "Climategate" as the latest sequel to the Nixon-era Watergate scandal, with leading climate-change scientists derisively labeled "warm-mongers."[60] A flurry (or dust storm, depending on whether you prefer your metaphors cold or hot within this context) of responses ensued. These revelations gratuitously added a torrent of jet fuel to an already-fiery dispute.[61]

When scientists and editors on either side of any contentious issue allow their own opinions to determine what will count as legitimate research and what will be worthy of publication, they violate the core principles of peer-reviewed scientific inquiry. If no one is "watching the watchmen," powerful people of either viewpoint can bar their "enemies" from access to publication in the very types of respected, prestigious journals that open doors to new ideas. If you believe, for example, that global warming is a fraud, and, lavishly bankrolled by major fossil-fuel industries, you control the peer-review system at one or more illustrious journals, you can exclude evidence of climate change from those important outlets, or attempt to overwhelm the truth with manufactured evidence of your own with ready access to other means of publication. Conversely, as in the case of the Climatic Research Unit debacle, maybe you can exert control over what gets into IPCC reports, and you can influence major journals not to accept articles that threaten the proposition of climate change as a genuine problem...literally forging a consensus.[62] In either case, this flagrantly corrupts what peer review is designed to achieve. Peer review becomes nothing but a stacked deck if all the "peers" are hand-picked to accept only one viewpoint and to "hide the decline" in temperatures indicated by some recent data. It is a politically-motivated *a priori* judgment on what ideas and data get through the door into the debating chamber rather than a careful, neutral, systematic, objective analysis of every competing theory's methodology, evidentiary support, and measurements.

Regardless of our motives, if we deliberately distort evidence, hide inconvenient data, destroy conflicting evidence, exclude and suppress opposing viewpoints, and generally collaborate to mislead decisionmakers, we do more than "merely" compromise our integrity and that of the scientific method. We forfeit our scientist's lab coat and trade it for a spy's trench coat. Incidents such as the Climatic Research Unit scandal take a maddeningly complex scientific challenge and make it enormously worse...while inflicting grave damage on the opinion many people have of scientists and the legitimacy of their theories. If scientists themselves lack sufficient faith in scientific inquiry to let their theories prevail or melt in the crucible of intense heat, why should laypersons trust the scientists?

Zealots can be their own cause's worst enemy. *Their flagrant excesses, when exposed, can turn impartial observers into skeptics even if the zealots' underlying concern is in fact well placed.* In their obsessive and excessive attempts to ensure the triumph of their view of the science, they undermine the credibility of even their own most fundamentally sound evidence. By deciding that they know best, that no one else can be trusted to reach the right conclusions, that no one else deserves to see all the evidence, and that the end justifies the means, they have cannibalized their own legitimacy. Jack Nicholson's famous line from the film *A Few Good Men* has been stolen by degree-laden people who should know better than to exclaim, "You can't handle the truth!" Once exposed as dishonest and manipulative in such matters, *all* their work, *all* their recommendations, and *all* their publications—irrespective of actual merits—will be considered suspect and unreliable by many less partisan observers. Ironic, isn't it? These are the tragic, almost mythical proportions of incalculable harm such misguided scientists inflict on the very ideas and ideals they claim to hold most dear.

The climate change debate—or war—or tag-team wrestling match—has evolved into something very different from the idealized concept of free, open, rational, methodical, fair, and objective scientific inquiry. The clashing titans taunt each other with thinly disguised variations on the venerable playground dialogue of "Is not!" and "Is too!" Like older, more degree-laden versions of street gangs, both factions flash intimidating signs and symbols at each other, but instead of complex combinations of digits on their hands, they use digits on handouts, tables, and charts. These wonkish gangs don't fight over street turf, but rather over who can count on a larger number of educated eggheads in the battle for who has the turf called "consensus" on their side. This risible rift finds the climate change activists claiming that "settled science" and consensus are on the side of catastrophic anthropogenic effects, and there's no time to waste in tackling the coming crisis. Meanwhile, just across the de-reasonableness zone, contrarians say that any consensus is actually in opposition to the alarmist view. Such dissenters collect thousands of skeptics' names for their "global warming petition project."[63]

What both sides seem to miss is that scientific consensus—*even when it does exist*—is no substitute for scientific evidence. Virtually all respectable scientists once believed that the Earth is flat, or that the Sun revolves around the Earth, but that didn't make it true. Ask crackpots and contrarians such as Kepler and Galileo. It took centuries of study, experimentation, discussion, analysis…and even courage…to derive fact-based, objective scientific truth instead of polling data. Today, it's true that no "legitimate" scientist questions the heliocentric nature of our solar system, but that only matters because the underlying science is theoretically sound, empirically verifiable, and sufficiently robust to withstand all competing views. That, and not unanimity, is what matters in science.

If we are truly interested in reaching the proper conclusions and decisions, and leading others in the correct direction, we are far more likely to persuade impartial and reasonable people if we remain scrupulously honest, objective, rigorous, fair, and committed to following the science wherever it leads. We would be well advised not to bar any person or any idea from publication without substantial credible evidence that this exclusion is warranted. As the saying goes, sunlight is the best disinfectant. If a theory, an idea, or a would-be scientific hypothesis lacks merit, the best way to establish that definitively and conclusively for all to see is to let it out in the open to compete vigorously with opposing views, in the bright light of *all* the evidence, in the vast marketplace of ideas. Let the erroneous theories fail, publicly, by fighting it out for all they're worth against better ideas and being exposed in their weakness, right there in front of everyone in broad daylight. The search for scientific truth is quite hard enough under even ideal conditions without letting a misguided let's-pretend game of "secret agent" derail the search for truth into a very public train wreck.

It would be a grave blunder to look aghast at the "Climategate" incident and conclude that there's nothing to the climate change threat. With excessive friends like those at the CRU, this issue doesn't need enemies; and some people who had already decided the threat was a hoax were only too happy to use this scandal as an excuse to declare victory and suggest we all go home. But every good cause, be it a religion, a nation, a form of government, or an environmental issue, has its extremists and fanatics. Bad causes have their extremists and fanatics too. You can't tell the legitimate causes from the meritless frauds by checking to see whether there are adherents who have gone too far. If "Climategate" teaches us anything, it is that we must be committed to full and open, methodologically sound, objective and impartial independent assessment of all the evidence from whatever source. The scandal doesn't end the debate. It only highlights the enormous importance of getting this question analyzed correctly, thoroughly, and fairly. This is no time to shut off the discourse and prematurely deem the case closed, much as some would prefer to do just that. Exactly the opposite is what's called for—the very type of vigorous, rigorous, open-ended and transparent inquiry that some people have been too afraid to allow. That's why I wrote this book.

Indeed, determining the truth with any reasonable degree of confidence and reliability, in something as incredibly elusive as the climate change context, is a daunting task under the best of circumstances. Let's get back to the business of figuring out what the science can actually show us, when we're not too busy fighting a disinformation war to consider the evidence. Not to belabor the obvious, but why should I stop now? A planet is a very big place, and it is not an easy thing to determine its temperature. For you or me, we might be able to afford to shed a few pounds, but we're of reasonably limited size. There aren't many options for measuring our temperature, and frankly I much prefer the mouth or ear methods to the main other alternative. But how do we take the Earth's temperature? It has no functional equivalent of our mouth, or our anus (unless you count the neighborhood where I grew up). So it is a matter of considerable scientific and technical difficulty even to choose where, how, and how often to take our readings. Our planet is, quite literally, a whole wide world, with many thousands of square miles of land and sea and a great array of diverse habitats and ecosystems. When the world encompasses regions that stretch from pole to icy pole, crossing forests, vast oceans, mountain ranges, grasslands, deserts, and equatorial rain forests on the way, it is a daunting challenge to arrive at an "average" or mean global temperature.

The methods we pick for conducting Earth's temperature measurements make a great deal of difference to the results we obtain. How many sampling sites should we use? Are the 517 weather stations currently in the global temperature-monitoring network enough, and are they sufficiently reliable or free from artificial perturbation?[64] How much and what type of supplemental data should come from satellites? How often should we take readings? At what times of day and monthly or seasonal junctures? What locations should we select?

At what elevations? Should the weather stations be allocated differently or concentrated in different areas? Should we take readings at the surface of some or all of the world's oceans, given that they cover most of the planet? Or perhaps some distance below the surface of the ocean waters? How many locations should we choose in polar regions, tropical rainforests, deserts, or temperate zones? How do we account for the temperature distortions caused by placing monitoring sites too close to major cities? Should we make an effort to pick evenly distributed measuring sites? How do we take into account the considerable practical challenges associated with placing monitoring stations and taking readings in the deadly cold of the polar regions, in the middle of vast rainforests, or atop treacherous mountain ranges? Can we conduct our readings remotely? If so, how should we ensure the equipment remains properly maintained and calibrated, even under extreme environmental conditions and with interference from local people and animals?

Our list of practical questions doesn't stop there. Should we concentrate our temperature readings on some regions instead of striving for roughly even distribution everywhere on Earth? Is there a good reason to consider some places more important, more reliable, more representative, or otherwise more appropriate that others as sampling centers? Should we try to weigh readings from some sites more heavily than others, maybe because they are more populous and/or more vulnerable to coastal flooding? Is there an advantage to be gained from extrapolating readings to fill in data gaps in certain areas, perhaps because these regions are isolated, remote, dangerous, unstable, or otherwise hard to sample directly? How can we compare readings from past years to today's data in light of major changes in the sampling methods, temperature measuring technology, number and location of measurement sites, and reliability of our information collection, storage, transmission, retrieval, and analysis processes and equipment? Should we try to correct or adjust the older data in any way based on recent refinements in the state of the art?

Okay, enough with the questions. Are there any answers? Well, let's see. Maybe we can begin to get a handle on the situation if we take an approach a little closer to bite-size than the entire globe. How about looking at just one city by way of illustration?

My home town, Chicago, Illinois, represents a good example on a much smaller scale of the challenges we face in determining Earth's overall temperature. There are great fluctuations in Chicago's temperature, on a daily, weekly, monthly, and yearly basis. These variations can exert tremendous influence on how we and other living things respond in our actual behavior to the temperature, at any point in time or in general. Chicago has an overall average annual temperature of 49° F. But that sneaky, boring, moderate-sounding 49° in no way conveys the extremes of both heat and cold residents have suffered there during brutally frigid winters and inhumanly hot summers.

Remember, another word for average is mean, and mean is an excellent description of Chicago weather at its agonizing worst. The average figures don't begin to reveal this crucial truth. For example, the average monthly temperature in Chicago in January is 21.0° F, and in July the average is 73.2° F. But the average *low* in January is 16° F and the average *high* in July is 84° F. And, of course, those long-term averages mask a lot of variation, with plenty of room for *much* higher or lower temperatures during any limited period of months or years, or at any given point in time. Take it from me. For people, animals, and plants actually attempting (somewhat foolishly) to survive outdoors even briefly in Chicago, these daily and periodic extremes may be more significant (and definitely more uncomfortable) than the harmless-sounding daily, seasonal, or annual mean.

Please understand that these difficulties apply even if we assume complete accuracy and reliability of our measuring equipment. If we manage to overcome the formidable obstacles complicating the design, production, protection, sensor function, data collection, data retrieval, and long-term maintenance of our array of tricked-out thermometers, we still must grapple with these sampling issues. And questions of where and when to measure temperature are inescapable, even on a very local level, and even today in our own backyard where we have every conceivable convenience and modern advantage.

Within any metropolitan area, temperatures tend to vary by several degrees depending on whether we take readings in the asphalt-and-concrete city center, the grass-and-trees suburbs, or the runaway-runway tic-tac-toe airport area. Various types of ground cover exert a major influence on the extent to which heat is collected or reflected, absorbed or dispersed. Chicago, as a city of considerable size, covers quite a bit of territory, and often noticeable differences in temperature exist depending on your proximity to Lake Michigan, the skyscraper forest of The Loop, or the outlying grassy suburbs.[65] And temperatures fluctuate further according to time of day, wind direction, and intensity of sunshine. Any average reading will be shifted by choices we make on when and where to take our temperature. So it makes a big difference for our Chicago "city temperature" readings, for any day, week, month, or year, just where and when and how often we capture our measurements. If this is true for the relatively compact geographical boundaries of any single city, it is true hundreds of thousands of times over on a global scale. Sampling decisions, in other words, greatly affect our results, even if we could somehow attain absolute perfection in our equipment's design and function. This doesn't make city or global temperature averages worthless, but it does require us to think about the choices, tradeoffs, and variations those averages tend to conceal.

All of these sampling variables become more elusive when we try to piece together a reasonable guesstimate of the Earth's overall temperature for times before the very recent past. Once we look back more than a few decades, actual temperature readings quickly become very scarce, haphazard, capriciously scattered around the globe, and unsystematic. And for the hundreds of millennia for which our only temperature data must necessarily be derived indirectly from scattered clues that happen to have been preserved in the fossil record, we are forced to take what little we can get. We deal with it because we have no choice, but our global temperature evidence from the long ago is extremely incomplete, unevenly distributed, and subject to varying expert interpretations. This is central to unlocking the presence or absence, as well as the size and direction, of very long-term trends in Earth's temperature. The extent to which contemporary climate patterns are a significant and worrisome departure from the historical context can't be determined with much confidence if the older information is too uncertain and unreliable to permit accurate comparison.

Expert-level brawls over such issues as the validity and meaning of the hockey-stick curve often stem in part from these complications associated with cross-centuries comparisons. Depending on our opinion of the quality and proper interpretation of well-aged temperature data, and on what data points we choose to include or exclude, we might develop vastly differing views as to the reality, magnitude, duration, and significance of the Medieval Warm Period. In turn, our perspective on the Medieval Warm Period and other possible candidates for "global warming" episodes from the distant past will be highly relevant to the way we think about current temperature data. If we view contemporary temperature numbers as fitting comfortably in line with other periods of similar or greater warming (or cooling) from which there were no major negative consequences, we will tend to have a rather calm take on the modern climate change controversy. This is acutely different from what we would feel if we consider recent trends as unprecedented and dramatically outside the historical norm. Such matters of evidence—what we accept or reject, what we deem significant or routine, how we place the present situation in broader longitudinal context—thus become considerations of paramount importance to our assessment of today's environment.

For any issue involving a question of severe harm from an environmental hazard, it is clearly relevant whether the hazard is unprecedented or has happened before. Anyone who wants to determine the degree of the current threat, the amount of damage we can realistically anticipate, or the nature and magnitude of the probable damage would want to know what came about during previous events of the same type. If the contemporary menace is truly unlike anything ever encountered, that is very significant. It means we will be in extremely unfamiliar territory as we try to assess the size and shape of the challenge, because it's without parallel--it's one of a kind, and we have no good point of comparison. Conversely, if the present-day situation does have strong similarities to prior events, the historical record from those antecedent versions will serve as a good starting point for figuring out what to expect this time around. In the climate change area, it matters a lot whether we have a reasonably similar context in which we can place our current situation. For example, if the Medieval Warm Period was a time of roughly equal or greater temperature increase compared to our present circumstances, we should be able to get a decent handle on what impacts we'll have to deal with, based on what happened back then. If ocean levels rose by a certain amount, or ice coverage receded to a particular extent, or the extinction rate spiked, we can use those effects as a beginning assumption on the probable impacts we will eventually experience in our era. But if we truly have no indication that anything like our current temperature trends has ever taken place in Earth's history, we will be forced to speculate to a much greater extent as to the meaning of our contemporary dangers because we'll have nothing equivalent to which we can compare our situation. The precedent, or lack of precedent, for today's trends is therefore a highly significant, yet uncertain, factor in our climate change decision equation.

The way in which we choose to handle these types of uncertainty, as that uncertainty relates to the momentous decisions we need to make, often is the single most crucial factor determining the outcome of any situation. It can literally make all the difference in the world. In the last few chapters of this book, I will focus on insights from some of the greatest thinkers in history with an eye toward making the best possible use of this uncertainty. Until then, let this little detour into some of the evidentiary complexities relevant to the climate change puzzle serve to highlight the pivotal, outcome-spinning importance of our efforts to make the right call in an atmosphere of doubt. Small wonder, with so much at stake, that people on both sides of the debate have so frequently resorted to personal attacks in lieu of rational, evidence-based deliberation.

The widespread human affinity for attacking one another aside, I think I've demonstrated that there are good and legitimate reasons for some of the factual and scientific uncertainties associated with the climate change debate. For any subject as vast and long-term as global climate change, there will always be difficulties inherent in collecting accurate data and correctly interpreting its meaning. We take the world as we find it, and the hard facts of life are that many places on our planet are such that conditions are harsh, access is challenging, equipment is difficult to monitor and maintain, and skilled observers are in short supply. Indeed, as we've seen, it is a formidable threshold problem even to decide what readings to include in our calculation of the global mean, in terms of the importance of attaining roughly even distribution across the spectrum of local circumstances and assuring the appropriate number and type of data points required. There is no consensus answer to that big challenge. But wait, it gets worse.

It is even a bigger challenge to produce planet-wide average temperature readings that the ordinary layperson (or the ordinary political leader, for that matter) will intuitively understand and accept as valid. This is a key point, because in the realm of law and politics (to the extent there's any difference between the two) it is not enough to have powerful evidence on your side. You also need to have evidence that the decision-makers will *perceive* as powerful. If the judge, or jury, or legislature, or citizenry is unable to understand the evidence, comprehend its significance, and weigh it within an appropriate comparative framework, the strongest evidence that ever existed will count for nothing. I'll return to this important but often-ignored burden-of-proof problem later, in Chapter 7. For now, just keep it in mind as an additional layer of difficulty looming over all the more specific details at issue in these grand-scale environmental dilemmas.

It is true that for many years dating back to the mid-19th Century, the samples used to piece together figures for average global temperatures were less than ideal. The thermometers from which regular data were obtained were not evenly distributed around the globe. They were over-represented in certain "First World" type nations and scarcely present at all in enormous territories throughout the oceans, as well as South America, Africa, and Antarctica. For periods more distant in time, our figures are necessarily derived from indirect indicators such as ice cores, tree rings, and corals. These are susceptible to varying interpretation, creating more room for doubt.

Similarly, with hundreds of thousands of years—and even millions of years—of Earth's climatic history as a backdrop, that's a whole lot of "context" ready and waiting for people to take things out of. To the extent that non-specialists grasp that the world is an exceedingly old place with a lengthy history of ancient and not-so-ancient shifts from ice age to nice age and back again, they may easily assume that our recent century-long upward creep in global mean temperature is well within historically normal limits. After all, as much as most of us love to complain about the weather, we are unlikely to bother uncorking any whine if the daily peak temperature in our neighborhood only oozes up about one single solitary degree, give or take, over the span of our entire lifetime. That minuscule, below-the-thermometer level of global warming is very easy to miss, given the usual much-larger temperature fluctuations we take in stride as we go from season to season and even from day to day.

People who learned everything they know about climate change from watching the popular film *The Day After Tomorrow* often have little trouble accepting the skeptical view of climate change. They have seen, or think they have seen, "real" climate change on the big screen, and it looked nothing like what they are seeing in the world around them. The movie depicted the ice age arriving at breakneck speed, with people fleeing down a hallway in a race to slam the door before the frigid wave of sub-zero temperatures caught them and instantly froze them. When millions of people see climate change dramatically represented as a fast-moving, even dizzyingly rapid phenomenon, they are understandably unimpressed by jargon-jammed, hyper-technical scientific reports of a slight bump in global temperatures during the past century. Their natural reaction echoes that of *Mad* magazine's fictional icon Alfred E. Neuman, who so often philosophized, "What, me worry?"

Of course, once we begin from the presumption that most people will not possess even the barest modicum of scientific savvy, it is difficult to fault the movie industry for perpetuating common misconceptions about climate change. Any motion picture that accurately depicted the realities of an on-coming ice age or tropical age would be exceedingly boring and well over the standard 2-hour running time. The term "gradual" doesn't begin to describe how climate change really happens. Even the most intense epochs of exceptionally hot or cold average global temperatures arrived and departed over the span of years, if not decades or even centuries. It didn't happen within a few seconds, as it does in popular movies, nor a few minutes, nor a few days, weeks, or months. You don't have to be a sprinter or even a crawler to be able to outrun an ice age that takes years rather than seconds to make it across a valley. That breed of invisible, virtually stationary, micro-incremental, interminably protracted transition in climatic conditions is not exciting, entertaining, or dramatic. Its only virtue is its truth.

But truth alone often hasn't been enough to bring about legal progress in environmental issues, and the global climate change question supplies evidence of this in spades. It is subtle, slow, and easy to overlook. We can't see irrefutable, unmistakable, shocking signs of danger, and thus we allow any uncertainty as to the reality, immediacy, characteristics, and magnitude of the problem to lull us into remaining comfortably nestled within the status quo. The prevailing paradigm holds that question marks trump exclamation points every time. So we keep our cool despite forecasts of meltdown and continue to sail effortlessly along on our carefree climate-change road trip, with our cruise control on and ourselves comfortably asleep at the wheel. But before we take a nap in a moving car, it's a good idea to make sure someone else is driving who is awake, responsible, and knows how. Maybe there really isn't anything to worry about here, but too many people have skipped directly to the nap stage without first verifying that the situation is truly under control.

On top of the scientific uncertainty as to the reality, dimensions, historical abnormality, and degree of danger tied up with the climate change dilemma, we also must consider the question of what we might be able to do to correct any such problem. It will surprise no one to hear that this is a difficult question. Obviously, when we dispute the very existence of the underlying problem, not to mention its magnitude, measure of certainty, extent of imminence, and other parameters, any possible cure will be at least as fraught with fuzziness as the disease. When we're talking about whether a course of action is reasonably called for (i.e., whether it's a good idea, all things considered), it makes a big difference whether the potential cure is relatively easy, inexpensive, equally shared by all concerned, reasonably sure to be effective, and free of major downsides, or whether it's onerous, highly costly, disproportionately burdensome with major free riders, of very uncertain efficacy, and comes with strings attached.

If climate change is happening, and if it's a major problem, and if it's being caused by net atmospheric increases in greenhouse gasses, are human beings capable of making a real and positive difference? If so, what are the most effective steps we could take? How much of the planet would have to be on board with the plan to make it reasonably effective? How certain are we of achieving the hoped-for results? How much money would it cost? What would it cost in quality of life and lost opportunities? And would it generate serious new problems of its own?

For a phenomenon as colossal as ominous shifts in the global climate, it isn't surprising that most rectifying proposals tend toward the mammoth rather than the minuscule. Some experts doubt whether human beings even have the capability to exert a meaningful and appropriately targeted influence on the planet's overall climate, irrespective of how many stops we pull out. They note that people, in the aggregate, have contributed only a portion of the total amount of greenhouse gasses in the atmosphere today, with other sources also adding appreciably to the always-present baseline quantities. There is also the challenge of lengthy lag times, with the prospect of many years between whatever actions we take and the actual appearance of positive results. It took centuries from the debut of the Industrial Revolution to puff up the problem to the level we now see, and there doesn't seem to be any good method of swiftly taking those levels back down, even if all people everywhere immediately ceased burning fossil fuels.

If we provisionally assume that, at best, we could only make a meaningful dent in the global climatic trend through the most widespread, giant-step, multi-faceted, all-hands-on-deck bailout über-project, some would question whether it's realistic to expect such unprecedented planet-wide cooperation. Would our notoriously selfish, parochial, narrow-minded, devious, competitive, and uncooperative species be willing to unite behind this common cause and collectively take one for the team? There have been some troubling hints that the neighborhood's big kids (nicknamed America, India, and China) would take their extra-heavy sacks of marbles and all go home rather than give up their gas-passing competitive advantage in the global economy. If any or all of these three "gas giants" (with apologies to the innocent planets Neptune, Uranus, Saturn, and Jupiter) refuse to play nice, could the rest of the world combined still suck it up sufficiently to make the necessary difference? Even if they could do so, would they, in light of the unfair economic edge this would concede to the stubborn bullies on the sidelines?

If we could somehow summon up the collective will, what would be the way? Proposals to ratchet down greenhouse gasses in the atmosphere usually focus on the obvious need to restrict or even eliminate the problem at its sources, which range from factory smokestacks to vehicle tailpipes to the tail-ends of domesticated livestock. Cap-and-trade plans would place maximum limits on industrial emissions of the regulated gasses, with a system put in place for entrepreneurs to buy, sell, and trade unused authorizations to emit. An initiative somewhat along these lines in Title V of the Clean Air Act has enjoyed some success, although that program is aimed at reducing other forms of air pollution and not greenhouse gasses. But how low would the caps have to be to produce the desired outcome, if a reduction in emission could work under any circumstances? And to what extent is this answer dependent on working out a solution to the free rider problem of non-participating gas giants? Are we in effect talking about the total abandonment of the internal-combustion engine and all other major uses of fossil fuels?

It's entirely self-evident that any action coming anywhere close to that level of systemic change would force what Al Gore years ago called a "wrenching transformation" of modern society.[66] It is difficult to conceive exactly what such a wrenchingly transformed society would look like, but it would require either a drastic overall reduction in energy usage or vastly increased reliance on non-fossil fuels, or both. We might try to return to something approaching a pre-industrial way of life, with all the myriad adjustments that implies. We could mandate much greater use of solar, wind, geothermal, and/or nuclear power, as well as other potential surrogates for fossil fuels. Or we could just take a deep breath, agree to presume that a considerably lower degree of wrenching might suffice, and settle for some compromise combination of conservation and alternative energy sources—and then hope that this is good enough to stave off the worst of the climate change calamities.

Similarly wrenching transformations might be imposed on other large sources of greenhouse gasses such as methane. The numberless herds and hordes of cattle, sheep, pigs, goats, chickens, turkeys, and other domesticated livestock we humans raise to feed ourselves and our pets are thought to be significant contributors to the influx in atmospheric greenhouse gasses. The incessant flatulence and belching from these tasty but tasteless and impolite masses can raise quite a stink in the global environment. Possible remedial measures span the spectrum from doing away with most of these gassy gastronomic delights once and for all to finding some less gas-provoking alternate food source or dietary supplement for them to eat—sort of like the animal husbandry equivalent of *Beano*.

In addition, we could try subtraction. We might attempt to augment the natural mechanisms that remove carbon dioxide from the atmosphere, i.e., photosynthetic plants, and greatly increase the amount of land devoted to tree and other large gas-guzzling forms of vegetation. If we succeeded in bringing about a large net increase in surface territory occupied by photosynthetic rather than just plain synthetic activity, we may be able to help draw down the aggregate amount of atmospheric carbon dioxide. In effect, this would entail an enormous push to reverse much of the staggering deforestation that's been inflicted on many of Earth's largest forests during the past few centuries. To have any real probability of making a significant contribution to aggregate atmospheric carbon dioxide proportions, this would need to be the most immense re-planting effort the world has ever known, with huge outlays of resources. Naturally, the land switched over to this use would no longer be available for industrial, residential, or animal-based enterprises anything close to the same extent as before.

I don't intend to paint an overly gloomy portrait of the available options to reverse climate change, nor do I want to deny the possibility of an unforeseen scientific or technological breakthrough at some point that would render remediation far more attainable at a much lower cost. This might in fact be an appropriate spot to dust off one of my favorite and oft-repeated lines from countless old monster movies: It's not much, but it's the only chance we've got. Still, if we want to retain our realistic, pragmatic perspective on the climate change question, we do need to remember that we have no experience at even attempting, let alone succeeding, to make a deliberate and purposefully directed modification of our planet's climate. On the contrary, we have legends such as that of King Canute (also spelled Cnut or Knut) the Great who supposedly tested his vast authority by having his throne set by the shore and then commanded the ocean's tides to halt and not dampen his royal robes or feet. The theory that the great king's power extended over the forces of Nature proved to be all wet. The futility of the king's majestic gesture, and his dampened enthusiasm for his own dominion, has been told and retold for nearly a thousand years as a vivid illustration of the limits of our human capabilities. As King Canute himself was said to remark, "How empty and worthless is the power of kings!"[67]

Of course, it doesn't necessarily follow that we remain powerless to stem the tide or cool the planet today, just because we've never been able to do it before. King Canute could not possibly have imagined the astonishing array of modern technological marvels we now have at our disposal as we contemplate staging a rematch of his legendary battle against the waves. Still, our only real track record to date on changing Earth's climate is entirely negative, accidental, and subject to varying interpretation. We may in fact have "succeeded," quite inadvertently, in turning up our world's master thermostat a bit after a couple of centuries of furious industrialization, urbanization, mechanization, and deforestation. In that lone example, however, we've had the distinct advantage of being motivated by 100-proof, Grade AAA, 24-carat self-interest as the high-octane fuel driving our hundreds of millions of independent bits of self-aggrandizing profit pursuit. We've also had hundreds of years to get the job done in our unknowing bid to (accidentally) turn up the heat. Yet even with all of that there remains debate as to whether our combined activities have really made a difference in the global temperature. Whether we are smart enough and/or strong enough now to deliberately do the reverse is just another big fat question mark to toss onto the heap we already have in our collection.

Near the end of this book, I will take on the dilemma of climate change more directly. For now, it's enough that we are aware of some of the notable gaps and questions in the factual and scientific foundation underlying the issue, as well as the types and magnitudes of changes we would likely need to undergo if we were to resolve to lower the planet's average temperature. It is important that we keep both of these ideas in mind—the uncertainties surrounding the reality, size, and imminence of the threat *and* the probable costs of taking the most likely set of corrective actions, assuming they could be effective. As we shall see, whenever we make any major decision involving considerable unknowns and significant costs, it would be an unwise if not reckless gamble to make that decision without giving appropriate weight to all the relevant risks and rewards, perils and prices, on both sides of the equation. Stay tuned until Chapter 8 and Chapter 9 for more about that.

For anyone who cares to look for it, there is a ready supply of evidence that can be assembled to make a case that global climate change is or is not a matter of great importance worthy of a massive mobilization effort to stem the tide. As I've indicated, the evidence tends to be complex, unfamiliar, counterintuitive, and difficult for a layperson to relate to. Some of it is scientifically suspect, but most people are unprepared to recognize this at all, or, if they do see the threshold issues, to evaluate rigorously the relative scientific strengths and weaknesses of the warring positions. The whole subject is light years removed from our normal field of experience, leaving us confused and highly dependent on the judgment of experts (real or imagined, hard-earned or self-proclaimed) to do most of the thinking for us. This is a perfect-storm environment for generating and perpetuating a great deal of uncertainty.

What is the real situation? Is climate change a disaster in the process of happening to us, or is it business as usual? Is this a risk that is worth making major changes in the way we live our lives and conduct our activities, or is it a mirage? Is it a case of people burying their guilty, greenhouse-gas-spewing heads in the sand and refusing to face the facts of the carnage they've unleashed, or a vast green-wing conspiracy of dishonest schemers who see the global warming bogeyman as their ticket to bigger, stronger, more centralized, more globalized government with themselves at the helm? Later in this book I will provide a rational, balanced, evidence-sensitive method of untangling this maddening mass of interwoven threads to find a way to deal appropriately with all that uncertainty.

Sadly, the climate change circular firing squad is not the only example of a dangerous and possibly endangered world paralyzed by uncertainty. Very much the same set of circumstances applies to the other prime example under consideration in this book: the failure of international environmental law. I refer here to the complete lack of an effective legal regimen to prevent or halt the current mass extinction…if there is in fact a current mass extinction. So fasten your seat belt. It's going to be a rough extinction-spasm of a ride.

CHAPTER 3

Modern Mass Extinction: Twilight of the Living Dead?

In this chapter, as with the previous one, I've deliberately chosen my objectives to take it easy on myself. I won't pretend that I've managed, in this single immortal chapter, to present to you in exhaustive and definitive detail all the information pertinent to the question of a current-day mass extinction crisis. The reality or fantasy of a vast modern die-off of endangered species to rival the dinosaurs' disappearing act has been literally dramatized to death in numerous well-researched and well-reasoned books (including two of my own making, about which more in just a minute, I assure you), and there's no possibility of adequately distilling their essence into these few pages. The evidence arguing for or against a contemporary genocide of genera and spoliation of species is vast and scientifically challenging to assess. Plus, there is much to discuss on (1) the various causes of any such extinction crisis, (2) the magnitude of the threat; (3) the imminence of the harm; and (4) the relative merits of an array of potential corrective measures we might take. I wouldn't want to burden you with all that heavy reading, and I surely don't want to burden myself with so much heavy lifting.

Well, that's a relief. I feel better already, don't you? My admittedly modest goals consist of establishing some credible evidence that: (1) there might (or might not) be a major contemporary emergency regarding the loss of biodiversity; (2) it should be of major concern to ordinary people everywhere if such a crisis exists; (3) the question is riddled with big, bad, obstinate unknowns; and (4) a human-generated campaign to ameliorate the situation to a significant degree would be possible but costly. After all that, I plan to leave you with two other questions: How bad is it? And is it worth the cost to us to do anything meaningful about it?

I know, I know. That's a lot of ink to spill with nothing much to show for it but a bunch of questions and no answers in sight. But that's what law professors do. We ask questions of our students to get them to think for themselves rather than spoon-feed them the answers. It's called the Socratic Method. If you don't like it, sue me.

I have written extensively about the contemporary extinction spasm and the abject poverty of the legal response. This is the subject of two of my previous books, *Ark of the Broken Covenant: Protecting the World's Biodiversity Hotspots*[1] and *Killing Our Oceans: Dealing with the Mass Extinction of Marine Life*.[2] I have also devoted several major law review articles to the twin crises of a present-day mass extinction both on land[3] and in the oceans.[4] Therefore I won't reiterate all of that detail here, but will touch on enough basic points to illustrate the similarity between this issue and that of global climate change. I want to show that, in both cases, we have a potentially devastating loss, plus some prospect of averting that loss through expensive and difficult intervention, but also some large and intractable gaps in the relevant information. So what's a planet to do?

As with climate change, the modern mass extinction is encumbered by the obstacles of scientific uncertainty and a worldwide rather than local impact.[5] Indeed, from the perspective of one reasonable frame of reference, there is even less hard evidence in support of a current extinction spasm. This view is predicated on the fact that a rather small number of species ("only" around 1,100 species by one estimate) are definitely known to have become extinct during the last several centuries.[6] According to the International Union for Conservation of Nature (IUCN), as of mid-2009 a total of 869 species were known to have slipped into extinction during the modern era, and perhaps another 290 were officially listed by the IUCN as "possibly extinct."[7] Compared to the over 1.7 million species currently identified as in existence on Earth, a mere 1,100 or so official extinctions is scarcely a blip on the radar screen—less than one tenth of one percent.

Try telling someone that we're aware of about 1,100 species that have died out during the last few centuries, but that if we subtract that figure from the 1.7-1.8 million we know are still around, we have over 99.9 percent that have not become extinct. Was their reaction a shriek or a shrug? How many times did you ever score 99.9 percent on an exam in school? For people inclined to limit their analysis to the most superficial level, that single fact essentially ends the debate, much as a deep local cold snap or severe blizzard is often accepted as conclusive evidence in refutation of global warming theory. Some laypersons will wonder what all the fuss is about, when the list of known recent extinctions is so short and we can still see tigers, giant pandas, and rhinos in the flesh (or fur or armor) at the zoo or on television.

It requires some comprehension of the nature of extinction, the gaping gaps in our knowledge of wildlife, and the factors that influence the scarcity and abundance of biodiversity to realize that actual, substantial evidence exists in support of a present-day mass extinction.[8] The IUCN's Red List of Threatened Species, as of late-2009, had some 17,291 species listed as "threatened with extinction" (defined by IUCN as falling within one of several categories of endangerment). This 17,291 species is many times larger than the number definitely known to have died out during the modern age, and could presage far greater extinction numbers to come if a sizable proportion of species currently on the road to extinction end up completing that sad journey.

But it could even be much worse than that. To put this in perspective, the IUCN is currently able to study 47,677 species, which sounds like a lot but amounts to only about 2.7 percent of all the species that we know for certain are now in existence. Moreover, the 17,291 "Red List" species list might be joined by many more imperiled species if more were known about life on Earth.[9] Even when scientists study nearly 50,000 species, as the IUCN now does, this number is dwarfed by the number known to exist, which is now somewhere between 1.7 million and 1.8 million species, the overwhelming majority of which have never been studied and might include many imperiled species that we're unaware of. Plus, the IUCN admits that the 47,677-plus species they're currently monitoring are not fairly representative of all life, nor do they adequately reflect what may be happening in many of the more remote and inhospitable regions of the planet.[10] Our biodiversity is so vast, so varied, and frequently so inconveniently or inaccessibly situated that we can scarcely hope to gain more than a fuzzy image of the true situation.[11]

Most people have no inkling that even the seemingly low number of verified extinctions recorded during the past few centuries represents a rate of extinction far, far above the normal or "background" pace of extinctions. Extinction is a natural part of life, to be sure, and even without any human involvement, species do tend to pass out of existence from time to time. The pace at which this happens varies a lot, by type of species, by habitat category, and by other factors, but there is a low background extinction rate that scientists usually consider the norm or the benchmark against which we should compare the pace of extinctions at any particular point in time. When species are dying out at a rate much higher than the background benchmark, that is a warning signal…even if the actual raw numbers still seem low to us.

So then, what is the benchmark, the norm, the usual background rate of extinction? Paleontologists generally agree that the background rate is between 1 and 10 species going extinct during any given ten-year period, per every 1 million species then in existence.[12]

This means that we can only compare current extinction figures with the very long-term background rate if we have a decent estimate of (1) how many species are going extinct anywhere in the world today per each decade; (2) how many species total are around at present; and (3) what was the most analogous background rate...perhaps closer to 1 than to 10 extinctions per million species per decade. This illustrates that there is a significant component of uncertainty even in assessing whether we're currently losing species faster than usual.

But not all species are equally at risk, and not all habitats have an equal number of species that call them home. On the contrary, much of the "diversity" in "biodiversity" might be understood as reflecting not just the great variety of life forms but also a diversity of numbers, degree of endangerment, and concentration in a few essential ecosystems. If one understands that hundreds of thousands of species are narrowly adapted to live only within certain specialized habitat niches, and that these species are incapable of adjusting to different conditions if their habitat is destroyed, the concept of endemism becomes clear. Endemic species live in one or more very particular habitats, and nowhere else in the world. Whether in certain types of coral reefs, or portions of the Amazon rainforest, or the hyper-heated waters of hydrothermal vents, or the vast open spaces of grassland, huge numbers of species are endemic to the *only* places suitable for their specific needs. If these species could live elsewhere and become more widely distributed, they surely would, but through many millennia of adaptation and migration, they have become irrevocably limited to one and only one set of ambient circumstances. If they do not have the proper range of temperatures, or type/amount of water, or variety of food, or protective shelter, or amount of sunlight, or any number of other environmental conditions, they cannot live there. That's what endemism means. A species can thrive beautifully within the habitat to which it is endemic, but be utterly unable to survive anywhere else on the planet. And when that one-and-only habitat is destroyed or significantly altered, a species endemic to it goes bye-bye. (You have my permission to tell this sad story to your kids as you struggle to persuade them to be less picky eaters. Good luck. Let me know if it works for you.)

Biodiversity studies show that endemic species are not in any way evenly distributed all around the world. Instead, they are heavily concentrated in a relatively small number of compact regions, both in terrestrial environments and in marine habitats. Comparatively limited geographical areas with a disproportionately large number of endemic species are sometimes called "biodiversity hotspots." On dry land, scientists have identified about 25 key hotspots that in the aggregate amount to less than two percent of the Earth's total terrestrial territory, yet are the only home for approximately half of all known species in existence.[13] With one in two known species worldwide endemic to these 25 hotspots, it should be apparent that tremendous numbers of species would die out if the hotspots were drastically reduced, fundamentally changed, or destroyed.[14]

This is in fact what has already been happening to the hotspots for many years. The terrestrial hotspots have lost about 90 percent of their primary vegetation since the beginning of the industrial age, and thus have been significantly altered and contracted if not entirely ruined.[15] But to appreciate what a 90 percent loss of usable habitat means to the endemic species that can only live in that habitat, it is necessary to understand how science extrapolates from one situation to another. Yes, this amounts to some detective work and connecting the dots…and it's a major source of the uncertainty associated with the issue of a possible contemporary mass extinction.

Islands, because they are literally surrounded by water and therefore present impassable barriers to migration in or out for many species, have been ideal natural laboratories for biodiversity studies. Isolated islands have furnished the basis for a scientific specialty known as island biogeography.[16] Studies on islands have indicated that, in general, a 90 percent reduction of available habitat will cause the eventual loss of approximately 50 percent of the species that are dependent on that habitat.[17] As the necessary habitat shrinks, the species that require it must crowd together into the remaining suitable area. They are forced to compete more aggressively with all the other species that also depend on that diminishing living space. The declining availability of sufficient food, water, room, shelter, shade, sunlight, and other factors eventually places lower limits than before on the number and variety of life forms that can survive there. And at some point, which differs for each species and for each unique situation, a number of the species on the island will no longer be able to exist there. They will cease to be present on the island—a phenomenon called local extinction.

Some of the biodiversity hotspots are in fact islands, or groups of islands. The other hotspots are—or at least arguably might be—the functional equivalent of islands, because they feature an unusual and specific combination of conditions ideal for the species endemic to each hotspot. The boundaries of each hotspot, according to the island biogeography theory, might as well be vast expanses of forbidding ocean insofar as the endemic species are concerned. They *must have* the particular set of circumstances found in their hotspot and, by definition, not available to an adequate extent outside their hotspot.

The hotspots are, in this view, ecological islands surrounded by other types of habitats unsuitable for the hotspots' endemic species. Thus, as with actual islands, if many, a few, or even one of a hotspot's defining characteristics are eliminated, at least some of the species endemic to that hotspot will in due course cease to exist. Islands of any type often are effective barriers to most species that might need to travel to or from the islands. They are surrounded by impassible and impenetrable walls, in effect, preventing migration in or out, and thus blocking both reinforcement and retreat. Many species that cannot fly or swim far enough, or otherwise survive the conditions beyond the local enclave, are prisoners there, with no possibility of parole or conjugal visits. That's what makes their survival so precarious. That's what makes the preservation of the precise and specific set of necessary conditions within the island sanctuary so utterly indispensable for them.

Island-biogeography studies tell us that if the hotspots have lost *90 percent* of their primary vegetation, which they have, this will result in *around half* of the species endemic to those hotspots going extinct eventually as a result of this habitat loss. The graph below illustrates this in terms of the number of species found on islands of varying sizes.[18] This is the result generally observed with endemic species on actual ocean-surrounded islands when their vegetation and other environmental conditions badly erode. When an island is defoliated by 90 percent, 50 percent of its species follow suit and disappear after some time lag. Roughly the same is *theoretically* the case whether the island is isolated from other suitable habitats by water or by other different and forbidding environmental conditions. With the terrestrial hotspots known to be the only habitat acceptable for around half of the identified species on Earth, that *could* spell the extinction of about 25 percent of all species now in existence...a mass extinction to rival the Big Five extinction spasms of the distant past. But again, this conclusion can't be reached by just glancing out the window and seeing shocking examples of bloody carnage all around. No, you can only get to an alarming mass extinction by looking long and hard in a lot of far-away places. It requires massive and sophisticated detective work, painstakingly gathering and carefully piecing together a collection of subtle clues that would give Sherlock Holmes a migraine. The evidence might end up spelling "mass extinction" to the educated expert observer, but the process we must use to get to that conclusion is anything but as simple as "A, B, C."

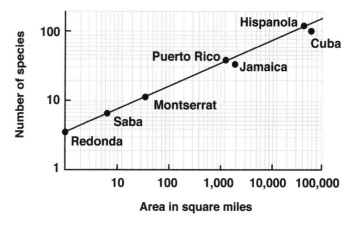

Number of species found on islands of different sizes

If we accept our imaginary detective's working hypothesis that key habitats are vanishing, and that it's mathematically predictable that this quantifiable degree of habitat loss will eventually translate into proportional species loss, that's very bad news. But it *might* even be much worse than that. The reason is that the hotspots are very likely the only habitat for many species that humans have never even identified. New species are discovered frequently, both in marine and terrestrial settings, and they are disproportionately found in the same areas of rich biodiversity as the previously-known species. This is not surprising, because we would expect on the basis of common sense that unidentified species would often thrive under the same combinations of habitat conditions that known species have found most suitable. If certain habitats, such as coral reefs and tropical rainforests, have proved to be especially accommodating homes for a disproportionately large share of all identified species, there is no reason to suppose that they would be any less suitable for the majority of still-undiscovered species. What's good for the goose is dandy for the dodo, or the dodo's unknown surviving relative.

Here is how the argument goes, in favor of the existence of large numbers of undiscovered species within the world's hotspots. The same characteristics that render these habitats ideal for known species (abundant food and clean water, reliably extreme-free temperatures all year, relative freedom from human interference, fertile soil, diverse and niche-rich terrain, etc.) would also make them excellent neighborhoods for life forms that have stayed under the human radarscopes. Plus, many of the hotspots are relatively remote from major centers of human habitation, whether in terms of sheer distance or difficulty of exploration or both. Thus they are less amenable to prolonged and intensive scientific study. When fewer knowledgeable specialists are consistently present on site to observe and collect, naturally less opportunity exists for people to encounter and to recognize species previously unknown to science.

Subtle distinctions between closely related organisms are often difficult to detect even under optimal conditions in the laboratory. So they become orders of magnitude even more elusive when researchers have to venture deep into unfamiliar and human-inhospitable territory. When species are differentiated on the basis of phenotypic traits as obscure as the number and arrangement of tiny hairs on a fly's back, it's all too easy to mistake numerous different species for just one species during brief, infrequent, dangerous, and unpleasant expeditions into the more distant reaches of Papua New Guinea, Madagascar, Indonesia, or Brazil. Trust me. I've been there. You don't know what uncertainty is until you've tried to count microscopic hairs on a tiny fly's backside in 100-degree heat and near-100-percent humidity in a jungle clearing with hot sweat pouring into your eyes. Under those conditions, better scientists than I have concluded that one fly looks pretty much like another and headed for the nearest air-conditioned bar. All of this means that the hotspots may well be the only home for half, or even far more than half, of the *unknown* living species, which have never been given a scientific name by human beings.

How many of these unknown species are on Earth right now, still awaiting human discovery? There is enormous scientific uncertainty on this key point. Estimates from well-credentialed experts vary so much that the entire enterprise looks like a guessing game.[19] Who can conclusively say whether any estimate (or wild guess) is right or wrong? I've often thought that an ideal job for myself would be "futurist," because my work would be error-free as far as anyone could know. No matter what wonders or terrors I might predict for the world of the future, no one could prove me wrong. Has anyone really come back from the future with proof that any of my predictions are off base? Of course not. Now that's job security.

In the same way, and quite naturally, no one can know definitively how many species still await discovery, because if we knew that we would have already discovered them. We do know that on Earth today roughly 1.7 to 1.8 million different species currently exist that proudly (or, to be honest, unknowingly) sport an official, human-generated scientific name. About one million of these are insects, and many others are tiny, visually unimpressive, decidedly uncuddly creatures such as mites, nematodes, annelids, and other unobtrusive living things. Are the 1.7 million currently identified species living alongside an equal number of anonymous relatives? Are there twice as many unknown species lurking out there? Could there be 10 million or more unidentified species on Earth right now? The answer has obvious importance for any question involving the magnitude of the current extinction problem, and, just as obvious, the answer is a gigantic question mark. It's easy to imagine vast multitudes of small, quiet, drab, non-threatening, physically unremarkable species leading publicity-free lives in many of the ecologically richest but least thoroughly studied habitats on land and in the oceans. There is simply no way to know how many are really there. The paparazzi are nowhere to be found when you really need them.

Many people would be surprised to hear any significant doubt remains as to how many different kinds of living things are around today. They would assume that some army of scientists is deployed all over the world, carefully studying all manner of wildlife, and has found just about everything there is to find. This viewpoint actually has some validity if one thinks of wildlife along the lines of the animals with which most people deal on a regular basis in their own lives. For many of us, especially within an urban setting, rarely see—or, more accurately, we rarely pay attention to—any non-insect species that isn't a cat, dog, squirrel, or pigeon. If that's all you have room for on your Ark, then you're probably correct in your assumption that few if any undiscovered species of cat, dog, squirrel, or pigeon are hiding somewhere in the world unbeknownst to humankind. Once in a while, there is a discovery of a new moderately-large mammal or bird; but for the most part we likely know about the vast majority of the big, obvious species on Earth today. They aren't small enough to stay undiscovered, for the most part and in most locations, with the possibly exception of some of the densest jungles and forests of Amazonia, equatorial Africa, and Southeast Asia. But we don't think about the types of living things we don't see or hear about.

Consider the one million species of insects here on Earth right now that actually have been given the *Homo sapiens* seal of approval in the form of a scientific name. When was the last time you noticed any insect that wasn't a mosquito sucking your blood, a cockroach scurrying under your kitchen cabinets, a fly buzzing around your head, or an ant crawling across your picnic lunch? Yet there are literally thousands—and in some instances hundreds of thousands—of separate and distinct *species* of such insects as beetles, moths, flies, ants, wasps, butterflies, bees, grasshoppers, and many others. We start with the same paradigm as in the situation with mammals and birds. The physically largest and most visually impressive types, particularly if they live in or near areas of significant human civilization, have mostly been found by some researcher and officially described.

But as we shift our focus toward the smaller, less colorful, and more remotely situated varieties, it becomes far less probable that someone has discovered them.

This principle grows progressively more powerful as the degree of difficulty inherent in reaching and seeing the species rises. Many insects are nocturnal, only becoming active under the shroud of darkness. Many thrive underground, burrowing and tunneling beneath our feet their entire lives. Many others are aquatic, living all or part of their life cycle under the water and away from our usual field of vision. Still others live inside other living things, including plants, mammals, birds, and other insects, out of sight. And all of these complicating factors become exponentially more intense as the entire inquiry shifts to far-off, inconvenient, uncomfortable, or downright dangerous regions. Tropical rainforests and remote island jungles are challenging enough to explore without adding in the extra obstacles that follow from the need to search in their dark, subterraneous, parasitic, and/or aquatic environments. And all the while we are looking for tiny, well-camouflaged creatures that often can only be distinguished from one another after detailed examination under a microscope of minute spines, hairs, and exoskeleton features.

If you were somehow to find yourself in a mythical insect museum where every known species is collected and preserved and on display in cases, you would find an astonishing number of superficially very similar species. Side by side by side with their nearest relatives, most of us would be absolutely incapable of detecting any difference between the species in row upon row of insects. Imagine for a moment how it would look. Imagine both the mind-boggling numbers and the mind-numbing similarity of case after case filled with the 300,000 different species of beetles (Order Coleoptera) currently known to exist. Picture the same cases packed with the 142,000 species of moths, skippers, and butterflies (Order Lepidoptera), 130,000 species of bees, wasps, and ants (Order Hymenoptera), and 100,000 species of flies (Order Diptera) humans have identified in the world of today. It would be an awesome sight in the traditional sense of the word. Personally, I would also be in awe of the people who devoted themselves to identifying all those myriad life forms...and of the thought that there could be so many more species like the ones represented by the specimens in the cases, still waiting for someone (not me!) to discover them.

This discussion should give you some appreciation of why there is so much uncertainty as to the number of insect species now in existence. But consider this as well: For many laypersons, insects are far better known and more often encountered and recognized than other highly populous categories of living things that are also good candidates for undiscovered species. We may not even know that there are such things as nematodes, for example. Yet scientists have listed thousands, and sometimes many thousands, of species of nematodes, annelids, mites, spiders, fungi, ticks, and countless varieties of plants...not to mention the dizzying profusion of species of single-celled organisms such as bacteria or archaeans. When we don't even know many of these entire *categories* of things exist at all, how would we recognize individual species of them and understand what they are, even if we should happen to come across them in the course of our activities? Answer: We wouldn't. And neither would almost anyone else.

Despite the astonishing diversity among the members of the Class Insecta, it may be that the greatest number of unidentified species are not insects but rather various types of marine life forms, separated from would-be human observers by that most formidable of barriers: oceans of water. Even more than the nocturnal life or the most remote and forbidding rainforests, thousands of feet of sea water in the vastness of the world's oceans forge a nearly-impenetrable screen concealing...who knows what. If people are able to visit these marine areas at all, by necessity it will be very briefly and under exceedingly challenging circumstances. Particularly in regions far from shore and miles beneath the water's surface, humans are at most just passing through. Even unmanned, remotely controlled submersible mechanisms cannot venture into many of the deepest reaches of the marine abyss. What species live there, and what manner of creatures are they? This is the unknown's unknown.

The vertiginous magnitude of the oceans hides a profusion of diverse habitats, some of which are extraordinarily inaccessible to human observers. Some areas are too deep—even miles and miles deep—for us to penetrate at all. Others, such as hydrothermal vents, feature superheated water so hot that it would be steam at ordinary pressures, yet they are home to some of Earth's most bizarre species. At the other extreme of temperature, portions of the deep ocean are dangerously cold, as well as subject to bone-mashing water pressure and unconquerable, perpetual darkness. Because of these amazing barriers to human exploration, entire major taxonomic groups like the Archaea were totally unknown to science until just a few years ago, and yet are now considered by some reputable scientists to be worthy of their own kingdom, alongside the traditional kingdoms of Plantae, Animalia, Fungi, Monera, and Protista. In a world where it is possible to discover not just a new species, and not just a new genus, family, order, class, or phylum, but a whole new *kingdom* concealed within the oceanic depths, it is hard to place any realistic limits on the potential vastness of marine biodiversity. It may not be literally true that anything is possible, but something close to it may well be the case.

This enormous uncertainty on a key factor as rock-bottom fundamental as how many species we're starting out with on Earth today makes the rest of the mass extinction issue incredibly speculative. Conceptually, it matters very much if we are at risk of losing 1,000 species out of a total of 1,800,000, or 1,000,000 species from a total of 3,000,000. It might seem strange to think of such a staggering quantity of ignorance—even about the most basic stuff of life—persisting in the modern world today, after centuries of scientific study and advancement, but it is very much the reality. The graph below depicts one interpretation of the available evidence, but as with many other such assessments, it represents extensive assumptions, extrapolations, and detective work.[20] Layer upon layer of estimates, analogies, and educated guesswork makes it exceedingly challenging to persuade decision-makers that there is a problem, and that it is worthy of their intervention.

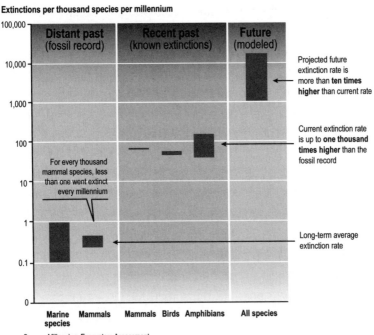

Source: Millennium Ecosystem Assessment

We have seen significant uncertainty as to (1) the magnitude of the extinction risk, i.e., the likelihood that impactfully large numbers of species currently threatened with extinction will in fact disappear within the foreseeable future; and (2) the total number of species now in existence that are endangered, both in absolute terms and as a percentage of all species currently on Earth. But even if rational people can accept the doubt and imprecision wrapped around these two points, we've still more question marks to deal with. One is the very mundane but natural tendency for people to wonder, to put it bluntly, "What's in it for me? Why should I care?" Phrased another way, the uncertainty centers on the degree of importance of the species at risk, or what it would cost us if we lost these species.

This is actually a deeper and more multi-faceted question than it might at first appear to be. It requires us to delve into fundamental issues of morality as well as the myriad ways in which other living things affect the health of this planet generally and human well-being specifically. I will try to touch lightly on the primary aspects of this highly nuanced puzzle.

At the most basic level, I believe there's no reason to dig for utilitarian rationales in support of preserving endangered species. For much the same reason why we refrain from needlessly harming other creatures in our everyday lives, we should just know it's morally wrong to drive entire species into extinction. It may be currently out of fashion or intellectually uncool to acknowledge the validity of such an ancient notion as absolute right and wrong, but that doesn't make it false. Even if it doesn't literally cost us anything in terms of our physical health or economic prosperity, don't we all implicitly accept the reality of morality and immorality? If you don't think so, let's try a little thought experiment.

Imagine yourself strolling alone down the sidewalk early one morning, carrying a wooden walking stick, with no other people anywhere in sight. What would you do if you saw a small puppy sitting quietly on the path right in front of you, with its little tail wagging and its bright eyes looking up at you? Be honest, now. As you consider your range of likely actions, does it even cross your mind that you might take advantage of this opportunity to use your walking stick as a club and kill that helpless, tiny pup? Certainly you would be able to do so, without any realistic fear of physical harm to yourself from the little puppy, and without any apparent chance that some observer would suddenly appear and either beat you up or report you to the police to be punished under the criminal justice system. So why wouldn't you gratuitously harm this defenseless creature that has so conveniently happened to be within your powerful reach? Would your decision change if, instead of a puppy, you encountered a robin, or a rabbit, or a butterfly, or a turtle? It's still a risk-free, no-strings-attached target of opportunity in every case. It's still a situation where the living thing poses no danger to you, constitutes no detriment to you greater than the minor inconvenience involved in stepping around it on the sidewalk; and it's not a potential source of life-sustaining food or warmth necessary for your own well-being. But why not activate your vastly superior strength and tool-using capability to snuff out any handy "lower" form of life of no obvious practical value to you, just for the fun of it, and just because you can?

We control ourselves because it's the *right* thing to do. What's more, it would be wrong to do the opposite. It's as simple as that. Intuitively, we know this. We might be burdened with years of higher education, and we might have been thoroughly marinated in the bland flavor of moral relativism, but at our core we are still human beings with an innate sense of right and wrong. No amount of education, or re-education, can completely eradicate our fundamental decency. It's wrong to harm other living creatures just because we can, and just because we don't see any particular value-added benefit to us from their continued existence. We get it. We would step around the small, living obstacle on the sidewalk, and maybe even pause to admire it or reach out to it gently, but we would "know better" than to club it to death.

On a far larger scale, the same ethical foundation applies when we have a choice regarding the extinction of entire species. If we can select among various reasonably-feasible courses of action, under any ordinary set of circumstances we should avoid wiping out a harmless species. This is one of those often-inconvenient "should" situations entailing a moral, ethical principle—a matter of doing what's right and refraining from doing what's wrong. Just as in our personal, individual example, we should not destroy another living thing—whether a single creature or a whole species full of similar creatures—unless necessary to serve demonstrably weightier purposes. Need I say more?

Probably. Even though we all accept the basic morality inherent in kindness (or at least absence of outright cruelty) to animals in our daily activities, that lesson sometimes gets lost once we move from the particular, familiar, and visible to the general, foreign, and unseen. People who would shudder at the thought of personally killing a cute, helpless puppy feel much less invested in the situation when we transform the question to the survival or extinction of entire species of beetles, nematodes, mites, archaeans, or annelids half a world away. Now we are no longer face-to-face with the potential victim, or even anywhere near it. We don't have to do anything ourselves—no physical action, no personal decision—to make the killing happen. We won't need to see the dead carcass afterward, or hear the sounds of life being extinguished. And we don't feel a direct, experience-forged, emotional link to the creatures being destroyed. We don't tend to identify with—or own pets that are examples of—such living, breathing organisms as beetles, nematodes, mites, archaeans, or annelids. It would be a rare person indeed who ever bent over and tickled a nematode while posing the oft-repeated rhetorical question usually reserved for our canine friends, "Who's a good boy?"

Objectively, all of these would be in the category of distinctions without a difference, and our correct ethical choice would remain to preserve rather than obliterate life. But realistically, these distinctions often do make a difference. Our history proves that they do. People who would never harm a furry little tail-wagging puppy wiggling happily in their hands don't perceive the core similarities to deciding whether to eliminate vital wild habitats in far-off tropical nations. Despite the fact that the habitat destruction would kill incalculably more creatures than just one single puppy, the two situations just *feel* so tremendously different that the decision-making process often yields very divergent results. Given that our response to an ethical/moral dilemma depends heavily on our affective, emotional involvement and connection to the situation, it is therefore necessary to look beyond bare invocations of right and wrong to address the question of whether we should preserve endangered species. More than just a matter of making the good or noble rather than the evil or destructive choice, the query, "To be, or not to be" thus also comes to resemble a calculating, cost-benefit, accountant's balance-sheet approach to matters of life and death.

Within this utilitarian, pragmatic paradigm, then, what are the uncertainties that cloud the value of living things? In what ways are species other than our own species worth saving? And how can we gauge the relative merits of saving or exterminating various species in various circumstances? Let's look at this momentous matter in the most no-nonsense, practical way possible...the actual cash-money value of species to us. We will begin with the conspicuous examples.

The economically valuable species that come most obviously to mind are the crop plants and domesticated livestock used primarily as food for humans and their pets. These plants and animals are actively protected (at least up to the point of harvest) and nurtured with as many nutritional supplements as we consider profitable. We usually raise such (literally) cash cows in great number and in huge aggregations, for reasons of efficiency and economies of scale. Suffice it to say we assure no appreciable risk of extinction for bottom-line important plant species such as domesticated varieties of corn, wheat, rice, oats, soybeans, and barley, or for their Kingdom Animalia counterparts such as cattle, pigs, chickens, and turkeys. Everyone fully understands the value of keeping these species around, and in vast numbers. We go to a lot of trouble to ensure that plenty of them reproduce and grow to the finest artificially-induced natural specifications money can profitably buy.

Things become a bit more complicated once we turn from species affirmatively cultivated, farmed, or reared by humans. What about seafood and other fish species, many of which are not raised by people but rather are, so to speak, hunted—fished, trapped, or otherwise caught in the wild? Laypersons certainly are interested in maintaining a steady, dependable supply of all their favorite varieties of seafood, and at an affordable price. These species include both vertebrates (all types of fish, including freshwater varieties such as trout and catfish and saltwater types such as tuna, shark, and swordfish) and invertebrates (shrimp, lobster, crab, octopus, squid, crayfish, etc.).

To the extent people think about it at all, they likely would favor some restrictions on direct over-fishing and habitat destruction, at least as much as is necessary to keep the prices low and the availability high. But in reality, this entails less-obvious but essential protection for marine sanctuaries, reducing accidental destruction of the other species relied on by the economically desirable varieties for their food, and restricting water pollution that could decimate both the target species and the ones that support them. Now the inquiry is migrating into areas with significant imprecision and unknown factors, because these free-ranging marine species can be highly mobile, hard to find, and dependent on food sources and habitat conditions poorly understood by their human consumers. It might actually be necessary to provide effective legal protection for more than just the "ka-ching, ka-ching" species if we want to keep the cash registers ringing at the supermarket and the restaurant.

It's possible that our tastes may evolve over time and we may someday want to grow or catch large quantities of other species beyond the relatively few we now use as our staple foodstuffs. Depending upon the culture and country, people are now eating some varieties of fish and meat that were not on the menu a decade or two earlier. This can happen by necessity, as previously desired fish species are over-harvested and driven into scarcity, and other types of fish we once disdained as junk are now proudly featured on restaurant menus. When you make a habit of eating your favorite seafood into oblivion and you still have a taste for more, it's easy for humble fish to make the transition from chum to yum. This dietary evolution can also transpire due to shifting trends and fads in human appetites, including the quest for more healthful types of food. People today consume meat from bison, emus, ostriches, and other creatures that were not candidates for the dinner table a short time ago.

Could it be that even more exotic species exist that we might eventually want to add to our diet? Whether or not we really want to, might it be that we'll have to broaden our nutritional spectrum in a world less amenable to widespread mass cultivation and/or more overpopulated with hungry people and pets? If so, there is no way of knowing for certain which species of plants and animals we could need to have available as sources of food. After all, back in the very distant past, who would have thought, based on outward appearances, that sophisticated and prosperous people would be eager to dine on such bizarre and downright creepy-looking creatures as lobsters and crabs? Invertebrate pot luck, anyone? Come to think of it, as long as we're willing to munch on members of the Phylum Arthropoda, there's very little fundamental difference between lobsters or crabs (Class Crustacea) and their cousins such as cockroaches, grasshoppers, and large beetles (Class Insecta). They're all arthropods, so why not? People already eat various insect species (in both the adult and larval stages) in some cultures, and this culinary practice could spread if conditions warrant it. When in doubt, it's arguably a good idea to keep as many potential food species off the extinction list as possible. Tastes change, needs change, and conditions change. A lot of Americans from just a few decades ago would be astounded if they could see their grandchildren today dining on something called sushi. The amount and type of food that we will prefer or even require a century or two from now may be profoundly and unpredictably different from the status quo.

Another major practical benefit that human beings derive from other species is their medicinal contribution. Many species currently play a significant role in either furnishing or facilitating the production of antibiotics, vaccines, pain relievers, healing agents, symptom-specific remedies, anti-inflammatory products, antidotes, and myriad other drugs. Some of these are enormously important life-saving medicines, while others greatly improve the quality of human life. The list of species indispensable to medicine is far too long to be included here, but it is loaded with obscure, seemingly unremarkable species from some of the most remote reaches of the Earth.

If we were to rely on a panel of experts to perform a global form of biodiversity triage, deciding which species to save and which to allow to slide into extinction, would we succeed in preserving the most medically essential ones? It's easy enough to make the correct call after the fact, that is, after someone has already studied a given species and determined its present medical usefulness and value. But how acute is our vision when we're trying to read the future?

Medically vital species don't come pre-labeled with an unmistakable sign reading, in many different languages, "DON'T KILL ME! I WILL SAVE LIVES!" Sometimes these living life-savers are disguised in the most baffling camouflage imaginable. Who would have guessed, in advance, that the humble, somewhat disgusting, blue-green mold known as *Penicillium notatum* would eventually produce a series of antibiotics capable of effectively treating a wide variety of deadly bacterial infections? Back in approximately 1870, even the finest medical and scientific minds in the world were unaware that such a phenomenon as antibiotics was even possible. Deaths from infection, resulting from both wartime and peacetime injuries, were a very frequent human tragedy of the highest order. There was no viable cure. Who could have supposed back then that (1) it was possible to develop an entire new category of medicines known as antibiotics that would vastly reduce the incidence of infection-related deaths; and (2) the revolutionary new medicine would trace its lineage directly to the *Penicillium notatum* mold? And yet, partially by the most serendipitous of accidents, Alexander Fleming and others would, within the next few decades, use that obscure mold to save countless human lives all over the globe. It's a good thing no "experts" had been there performing triage on which species stay or go away, or that yucky little mold wouldn't have been around for anyone to notice its strange tendency to inhibit bacterial growth.

To paraphrase Winston Churchill, never before have so many owed so much to so humble an organism. The penicillin example dramatically illustrates the insanity of relying on human judgment to decide which species may safely be shoved into the ovens of extinction. We're just not smart enough to know when we're wiping out the next medical breakthrough as we cavalierly dispose of all those tiny, ugly, uncuddly, uninspiring species. No one may ever name a sports team after it, or make a Disney movie about it, or use it as a model for a Webkinz stuffed animal, but *Penicillium notatum* and numberless, often nameless species like it may be among the most valuable species on the entire planet.

The uncertainty surrounding which species have current medicinal value becomes exponentially greater when we consider the prospect of new medical threats looming off in the future. We may not often think about it, but the number, severity, and diversity of medical hazards facing people on this planet varies immensely as time passes. Diseases arrive and depart, much like species themselves and the individuals that collectively comprise each species. There is absolutely no reason to believe that the world has already seen every specific form of a major medical threat—every type of epidemic, pandemic, or plague—it will ever have to endure. Precisely the opposite is almost certainly the case.

Even within recent history, important new diseases have burst onto the scene to exact a terrible harvest of human life. This list of unprecedented and unpredictable killers includes, of course, Acquired Immune Deficiency Syndrome, or AIDS. It's easy to forget, but AIDS wasn't even known to science until 1981, when the first few cases were identified in human beings. Yet during the first three decades of its awful existence, AIDS has exploded throughout the peoples of the world, sparing no nation or culture. During 2007 alone, AIDS took the lives of more than 2 million people worldwide, including some 330,000 children...all from a disease that didn't even officially exist in 1980.[21] This could happen again, at any time. We've no way to predict or prevent such dreadful onslaughts from medical intruders.

Viruses mutate and evolve, sometimes developing new and deadly hazards to human health. The same is true for disease-causing bacteria, which sometimes produce new strains resistant to our arsenal of antibiotics. At the microbial level, evolution can take place at dizzying velocities, with new generations adapting to changing conditions every day. Moreover, whenever there are new interactions between people and animals, or unprecedented migration into different habitats, or new uses for familiar species, or any change in the way disease vectors operate within and between species, the opportunity grows for creating novel medical dangers. And there is even the chance, albeit remote, that returning space probes, shuttles, or capsules might someday bring back from a visit to another planet the worst possible kind of souvenir—an entirely foreign form of disease-causing agent never before present on Earth. Any of these could either attack humans directly, or work their malevolent magic indirectly through an assault on the plant crops and livestock animals on which we depend for our food.

If (or, far more likely, when) any of these events come to pass, it may open the door for some previously "unimportant" species to serve as the source for the cure. Some distant cousin of the penicillin mold may be out there, just waiting for its turn to shine as it engages in mortal combat with the unwelcome new intruder bacterium. Or some genetic traits may be coded within the DNA strands of a tropical frog, shrub, monkey, weed, worm, or beetle that could produce a genetically-modified vaccine or antidote or disease-resistant strain that can withstand the new medical peril. Phenotypic (actual physical) features and genotypic (genetic code) traits can both exist for hundreds of thousands, even millions of years without demonstrating any particular utility. They are simply there, waiting on the shelf just in case, so to speak. As long as they are not actively harmful to the species carrying them, they can be perpetuated for innumerable generations, devoid of any identifiable purpose.

But once circumstances change, the species that carry these ready-made curative characteristics can be extremely handy to have around. And we can never know in advance which species might one day, under a changed set of circumstances, rise to the occasion like a latter-day *Penicillium notatum*. This optimistic scenario depends, of course, on our obscure little heroes-in-waiting not becoming extinct before we end up desperately needing them.

Before the next AIDS—or something even more deadly—emerges from the shadows and imperils us all, we are constantly haunted by a question that probably never occurs to most of us. The somber truth is that we have no reliable crystal ball to guarantee that we are, today, presciently preserving whatever unremarkable species will someday suddenly become the rescuer of millions of people. Are we extinguishing species right now that might have saved our own grandchildren from the rising generation's pandemic or plague? How much are we willing to wager that we're guessing right? How many lives are we ready to bet on our notoriously faulty powers of prediction?

Naturally (and I mean that literally), it may be that the actual particular species we need for any of the above services is itself dependent on the continued existence of numerous other species that serve no direct and obvious human purpose. Every species is part of a complex ecosystem, with an intricate food web and an interlocking series of symbiotic, parasitic, and predator-prey checks and balances. The loss of *any* member of such a labyrinthine living network can initiate a cascade of toppling dominoes, often in unpredictable ways, and can take with it other species vitally important to human health and prosperity. You can't safely think of any species as an island in isolation. The linkages between and among species are numerous, convoluted, and poorly understood. We typically have no way of knowing whether the extinction of any one species, however humble and obscure, might indirectly result in the loss of seemingly unrelated and far-removed species of great utilitarian value to us.

Is this enough uncertainty for you yet? Well, not so fast. As they used to say in a lot of annoying television commercials, "But wait, there's more!" I will touch on just one last example of how we cannot be certain of the practical value of any species. This may be more significant, in the aggregate, than all the other benefits we derive from other living things. I am referring here to the concept known as "ecosystem services."

In a stealthy manner that is often imperceptible to people, a host of organisms constantly engages in processes absolutely essential to the continuation of life on Earth. Most of us never realize this is happening, because it is all so ubiquitous, undramatic, and familiar. We take it for granted. But far more significant than the provision of any particular type of food or medicine, these ecosystem services form the mechanism that moves all of life.

You might think about chickens or cattle when somebody

mentions important species, but do you automatically focus on honeybees, sphinx moths, hummingbirds, bats, butterflies, wasps, or any of the other species that pollinate so many of our most valuable crop plants? Without these unobtrusive little legions providing pollination for thousands of vital plants all around the world, there would be a catastrophic loss of fruits, grains, and vegetables upon which people rely so heavily, whether directly (as our own food) or indirectly (as feed for our livestock). We possess no substitute. Even if we would never dream of making a meal out of any of the pollinators themselves—moth casserole being a distinctly acquired taste—we would soon be desperate for most of our familiar types of food if the pollinators were gone. Wind-borne pollination only works for a small fraction of plant species, and the rest of our cherished "star" species would crash without the supporting cast of the pollinators.

Likewise, it's understandable that the species mentioned affectionately in the lyrics to the famous song "My Favorite Things" from the Broadway musical *The Sound of Music* are roses, kittens, ponies, and geese. None of these beloved living treasures are involved in the decomposition of wastes—despite the fact that it is another crucial form of ecosystem services. Even a lyricist as gifted as Oscar Hammerstein II didn't make room among the list of favorites for such essential decomposers as annelids, grubs, beetles, bacteria, flesh-flies, fungi, or blow-flies. And yet the breakdown of dead, excreted, or discarded organic material, whether from plants or animals, is an extraordinarily significant part of our Earth's life cycle. They may not appear as beautiful as a rose or as cute as a kitten, but these decomposers are literally unsung heroes.

The many species responsible for waste decomposition clear the ground so that seeds, precipitation, and sunlight can get to where they need to go, nurturing the next generation of all types of plants and the animals that eat the plants. They often loosen and aerate the soil as well, further facilitating plant germination and growth. They reprocess what otherwise would be ever-higher mountains of detritus to make room for animals to move, breed, find shelter, and eat. And they recycle the chemical components of waste so that they can be used again to sustain new life. Some of these species also help to detoxify certain waste materials and prevent them from poisoning water supplies or otherwise spreading disease. It's not easy to imagine what would happen if someone or something disrupted the process of decomposition and detoxification, because these break-down species just keep going and going like an un-cute miniature army of Energizer Bunnies. But considering the massive quantity of leaves, feces, bones, feathers, carcasses, tree trunks, and dead plants added every day to the Earth's surface, it would be anything but a pretty sight. Not only would the world become much uglier, smellier, sicker, and messier, it would be utterly unlivable for the vast majority of all living things, favorites or not.

There are other equally indispensable ecosystem services as well, and these, too, are often provided by species we never notice. For example, all photosynthesizing (green) plants assist with carbon fixation or sequestration and its conversion to oxygen, giving us air we can breathe. From the tallest redwoods to the tiniest unnamed tropical flower, these members of Kingdom Plantae collectively remove huge volumes of greenhouse gasses from Earth's atmosphere 24/7/365.25 and in return produce oxygen for members of Kingdom Animalia to sustain our lives. This is two ecosystem services for the price of one, because plants simultaneously counteract the harmful climate-changing properties of excess greenhouse gasses (including those artificially added to the atmosphere by human activities) and refresh the air we (and all other mammals, birds, insects, reptiles, etc.) rely on for our respiration. Not bad for a Kingdom with hundreds of thousands of different species, yet not a single Major League Baseball team named after any one of them...not even the Chicago Onions or the New York City Apples.

By now I think you get the picture. Ecosystem services rock. And I haven't even mentioned the tremendous importance of pest control, or any variety of biological checks-and-balances, which supplies life-saving natural regulation in numberless ecosystems all over the world. Likewise, some of the humblest species play major roles in disease control and prevention, seed dispersal, and the various biogeochemical cycles (involving the capture, metabolism, and recycling, not only of carbon, but also nitrogen, oxygen, phosphorus, and water). Many of us never even realize that some of these ecosystem services exist, because they happen so unobtrusively, so reliably, and so deep behind the scenes that we can comfortably take them for granted.

Here's the biggest unknown involving ecosystem services: What would happen to them, and to us, if we lose large numbers of key service-providing species to a mass extinction. Much like the vital organs in our own bodies, the ecosystem service species perform their crucial functions transparently and without fanfare...if all is well. Each of the ecosystem service networks of species is like a person's heart. If it's working properly, you never know it's there. You only notice it when there's trouble. And if there's trouble, it can be very big trouble for a lot more of the body than just the heart.

I will mention one more factor that makes the mass extinction issue so troublesome for politicians and citizens who want to support political action grounded on firm facts and solid, if not irrefutable, evidence. That factor is the slowness of the extinction process in most situations. As with climate change, mass extinction is especially insidious and elusive of legal action because it happens so gradually. It's imperceptible to any ordinary observer.[22] It's the classic silent killer, and it kills so incrementally that many generations of human lives can come and go while watching in vain for the evidence of extinction.

Many people are completely unaware of the way extinction usually happens. They assume that it is a rapid process lending itself to accurate measurement and direct observation. And they often think that the usual primary cause of extinction is direct hunting by humans, as with the dodo or the passenger pigeon. Conditioned for fast-paced or even instantaneous results by such modern influences as television, movies, video games, Twitter, and the Internet, they expect extinction to follow the model depicted for climate change in *The Day After Tomorrow*. The presumption is that any extinction worthy of the adjective "mass" would have to be startling, sudden, apocalyptic, and overwhelmingly, inescapably horrifying.

The truth is far more subtle than the myth. The main cause of all the mass extinctions in our planet's history appears to be habitat destruction or modification, including the effects of climate change. This is the case even for global die-offs initially touched off by a one-time disaster along the lines of an immense meteor strike, such as the so-called K-T extinction spasm (at the boundary between the Cretaceous and Tertiary periods) about 65 million years ago that wiped out the dinosaurs. By far, the majority of the species lost were not eliminated by the original meteor crash itself, nor by the direct and immediate effects from the fires, tsunamis, and earthquakes spawned by the impact. Only when Earth's climate shifted from the colossal quantities of smoke and particulate matter unleashed into the atmosphere after the meteor struck did most of the species move toward extinction. And this long, protracted deathbed tragedy required centuries and even thousands of years to unfold fully. Again, it was not the jaw-dropping drama screenwriters seek out, at least not once the initial "big bang" was over. It was a brief moment of action, followed by millennia of agonizingly incremental after-effects.

The same prolonged process is typical of any extinction brought about by habitat modification and destruction. In areas such as today's biodiversity hotspots or island habitats, the loss of hospitable habitat does not take the entirety, or even a sizable portion, of its toll immediately. The last, full measure of extinction fallout from habitat loss can take decades, centuries, or millennia to arrive. This happens very gradually as available habitat is reduced, first in one sector, then in another, here a little, there a little. It unfolds like a titanic flag slowly unfurling as some species struggle to out-compete others for the ever-scarcer nutrients, water sources, protected alcoves, and other necessities of life. There usually are no instant winners or losers in this high-stakes game of *Survivor*. Most species can, so to speak, circle the wagons and pull back to a more limited area in the face of the growing competition for shrinking resources. They can retreat, little by little, and fall back into ever-diminishing territory, and move by fits and starts over many years into a form of "last stand" mode. This sad twilight fight for life can go on for much longer than any person's lifetime, and often goes completely unnoticed by human observers.

Eventually, as suitable habitat vanishes and competition intensifies, a species' numbers and range become eroded to such an extent that extinction is inevitable, and literally becomes just a matter of time. The scientific term for species in this situation is "committed to extinction," with the colloquial variant "the living dead."[23] Once a species declines to the stage where it is committed to extinction, it is, as a group, terminal; it can never recover to the point of indefinite sustainability.[24] To be sure, a species can hang on as part of the living dead for a very long time, sometimes for centuries or more, and thus appear to most observers as if in no particular danger. But a species committed to extinction has been irrevocably weakened, and sooner or later something will come along and deliver the ultimate coup de grace. It might be a wildfire, human-induced habitat modification, an earthquake, changes in climate, an epidemic, or the advent of an invasive species; but eventually the living dead will be pushed past the tipping point into actual and everlasting extinction. Again, the fact that a species can linger in this fatally debilitated state for hundreds of years is insidiously deceptive. It causes many of us to overlook the extinction process altogether, and to remain entirely unaware that species are invisibly slipping past the point of no return onto death row right before our eyes.

For most Americans, the evidence of our current mass extinction is at least as difficult to detect as that of climate change. One reason why: The bulk of the extinctions are taking place in the hotspots, almost all of which are either in other nations or underwater in the oceans. And, given the profound gradualism of the extinction process, it appears superficially that nothing is happening. When people ask me what a mass extinction looks like, I tell them to take a peek out the window. That's it. We may well be in the midst of a mass extinction right now, and yet from all we can observe in our own lives, everything is normal. Our mass extinction is doom disguised as boredom, or crisis camouflaged as stasis. Sadly, if the mass extinction is so imperceptible as to be boring or unreal to us, we are unlikely to take any steps to intervene, and extinctions will continue unchecked. We will deem the notion of a modern mass extinction as highly uncertain at best and unworthy of our attention. Hundreds of thousands of species will literally be bored to death.

And there is still another question mark to grapple with. As in the previous chapter dealing with climate change, we need to discuss the advantages and disadvantages of some of the most likely remedial measures. To put it another way, could we stop a mass extinction if we tried, and if so, how much would it cost? I can't adequately cover this vast topic here, but I do want to touch on some of the practicalities and some of the costs associated with a few of the corrective measures we might implement in our efforts to avert a potential mass extinction. Without at least some ballpark concept of the price tag attached to these well-intentioned potential interventions, we are unprepared to conduct a coherent and rational analysis of the basic question: Is it a good idea to bet on them? In other words, if we get serious about gambling on saving most or all of these species and their habitats, what would we have to give up in the bargain? I've tried to outline in some detail why there's reason to believe we have a lot to lose by doing nothing to save biodiversity. But we'd be betting in the dark unless we also have a fair idea how much it costs us to play, and what we are risking by getting involved, as well as what's at stake if we don't take the plunge.

As with climate change, the question of what we could do about it that would actually help is not an easy one. It comes with its own barnacle-like question marks attached beneath the surface. Part of this uncertainty stems from the presumption that there may be numerous species on Earth today still waiting to be discovered, so we have to deduce what needs these species might have and where they might be located. That's a lot of specificity to ask when we're dealing with hypothetical creatures no scientist has ever described. But we could reasonably proceed under the assumption that most unknown species would reside in the same biodiversity hotspots where we find the majority of known species, and that they would have roughly similar requirements for survival. If they didn't, why would they live there in the first place?

Our experience with domestic statutes like the Endangered Species Act and its international relative, the Convention on International Trade in Endangered Species (CITES) suggests that any attempt to safeguard individual species, one at a time, would be exorbitant and inefficient. The presence of a few endangered straggler species like the Red-Cockaded Woodpecker or the Northern Spotted Owl in areas of extensive human activity and habitation can lead to very high lost opportunity costs if we force people to shut down their productive activities and make their property safe for the creatures living in paltry numbers among them. Some imperiled species are considered keystone or indicator species, and can serve as surrogates for additional life forms in the same habitat with similar living requirements. But ordinary, isolated species are prohibitively resource-intensive when we try to rescue them one at a time from their decimated, deathbed predicament.

The evidently disproportionate cost of saving individual endangered and threatened species generates an additional consideration as well. Public opinion tends to mount against these bailouts of belly-ups when it seems unfair to force ordinary folks to give up their livelihood, their family farm, or their timber-cutting for the sake of a few owls or woodpeckers. People wonder why the government can compel average citizens to sacrifice so much for a few struggling birds, and yet pay the affected persons no compensation for their losses in return. At least with traditional government "takings" of private property for public uses (through eminent domain) the government pays the property owners some "just compensation" for the loss of their land.

In contrast, in the case of regulatory takings like those caused by the Endangered Species Act, the property remains technically owned by the private individual. Therefore, unless a court decides that the government's regulations have deprived the property of essentially *all* economic value, the action will not be considered a taking and no just compensation will be paid. Voters look at these situations and get angry. They become opponents of conservation measures, because they see them as unfair and unequally burdensome. When this happens over and over, public support for biodiversity preservation erodes badly and it becomes even more difficult to marshal resources for additional efforts.

The most effective, practical, and efficient corrective action for the mass extinction crisis would probably be very different from this pricey, ad hoc, and politically dubious individualized focus on saving one species at a time. Consistent with the hotspots concept, the optimal approach would seek to create and actively safeguard a carefully-selected network of nature preserves, wilderness areas, and wildlife refuges. These would encompass to the greatest extent feasible the habitats—that is, the habitats themselves—with the world's largest numbers of endemic species. Within the broad boundaries of this general idea, room remains for myriad adjustments and refinements to fit the situation's realities on the ground in any particular region. Many of the hotspots are not currently under formal legal protection at all, and others are only partially included under any existing program of legally recognized sanctuaries.

Whether officially called national parks, national forests, wildlife refuges, wildlife sanctuaries, wilderness regions, marine protected areas, regional set-asides, or any other similar name, these various legal measures have been woefully underemployed in serving the planet's richest pockets of endemic life. With such a high percentage of Earth's biodiversity concentrated in so few relatively small areas, there is room for the law to make a major difference. Nature has put many—maybe even most—of her eggs in just a small number of baskets, so if we can provide armed guards for those baskets we should be able to save lots of eggs in a cost-effective way. But this hasn't happened yet, and much of what has been done generally is worth little more than the paper it's printed on.

In terms of protected habitats, the options range from comparatively modest restrictions on habitat destruction or modification to virtually total bans on human intrusion. Within the American conservation system, national parks and national forests would be at the former end of the spectrum, while national wilderness areas are at the latter. Serious political and practical factors intertwine with decisions on what degree of human activity to allow in and near the protected areas, and tradeoffs making conservation more attainable and affordable often cannot guarantee as secure a living environment for the resident biodiversity. A "multiple use, sustained yield" juggling act approach is an easier sell politically and economically. But it may not provide adequate guarantees that the local ecosystems will remain intact in the midst of all the other functions burdening the habitat.

Depending on which centers of endemism we select for protection and how expansively we define their physical boundaries, millions of people will be living, and trying to make a living, within the edges of many of the hotspots. Although some hotspots—with approximate geographical limits as they are usually understood—are rather sparsely populated with human beings and are essentially wilderness regions already, other hotbeds of biodiversity are home to surprisingly large numbers of people. If we accept the proposition that human-caused habitat destruction and modification is the primary threat to biodiversity today, we might also accept this corollary ("Kunich's corollary" perhaps?) as an oversimplified but still useful rule of thumb: The fewer people who are present anywhere near the habitat, the more likely it becomes that any species within that habitat will survive.

As long as I'm oversimplifying, let me keep it going a bit longer. The truth at the core of Kunich's corollary is that, irrespective of our intentions, people generally tend to do more harm than good to wildlife. We as a species are highly gifted when it comes to killing and destroying, but considerably less adept at nurturing and preserving Nature's living resources. It's easier, quicker, and demands much less knowledge or understanding. When we decide to intervene on behalf of any wild creatures, our actions often quickly outpace the strict limits on our grasp of what makes Nature work. We just aren't smart enough, and we haven't done nearly enough homework, to overcome our innate and massive ignorance of other living things. Even with the best and noblest of intentions, we tend to do more damage than damage control. In light of this, we may often prove most effective at preserving any pocket of biodiversity when we resolve to keep people as far away from it as possible, to as great an extent as possible. It's a humbling thought—that we help the most when we butt out and keep our butts out of the way—but there's more than a grain of truth to it.

This hands-off and butts-out paradigm might generally become the most effective for preserving the maximum amount of biodiversity in a majority of cases. Such a method should be especially appropriate where we suspect that large numbers of unknown species, with unknown specific needs, are hiding undiscovered within the deepest reaches of a given life-laden habitat like the hotspots. In those black-box situations, where we really don't know what we don't know, the most generally applicable and broad-based preventative approach is probably the best we can muster. In effect, if you don't know how many eggs are in your basket, but it feels heavy, the wisest course of action seems to be to take very good care of the basket itself—doing all you can to make sure nothing smashes it or sets it on fire—and trust the inner details to Nature. But this strategy clearly comes with a higher price tag than less comprehensive and humans-keep-out conservation measures. It costs large sums of money to move people out and keep them out, and it generates a high price tag from *lost opportunities* to farm, mine, build, drill, travel, reside, and graze within the off-limits areas as well.

Any new system of legally protected habitats would need to be not only scientifically selected but also very well policed and monitored to prevent the rampant poaching, timber-cutting, capturing, and other illegal activities that have turned many existing preserves into little more than "paper parks."[25] Great big piles of money can be made in endangered species, whether dead or alive. With such large incentives spurring on the poachers and collectors, only a well-trained, adequately staffed, heavily-armed, and well-equipped force of dedicated professionals guarding the wildlife refuges can hope to prevent widespread devastation. A major park-ranger force of this type is hard to assemble under any circumstances, given the grave dangers involved and the immense incentives to join the enemy side, but at a minimum it would be very expensive. It also costs money to build and maintain the fences and other barriers that might keep wildlife in and hunters out. But again, the greatest cost of a comprehensive hotspots conservation network is in the foregone economically profitable opportunities.

Many ways abound to tailor this type of conservation program, but all of them would entail strict limitations on any new or expanded human-related activity likely to harm the protected area. Any push to freeze the level of human intrusion at no greater than its current dimensions would necessarily mean surrendering the opportunity to make further inroads into the habitats. What dollar figure could we place on giving up the chance to start new mining operations or enlarge existing mines? How could we price out a ban on new oil and gas exploration and extraction, or timber-cutting in the vast rainforests? We can't possibly know for sure how much oil or gas we might have discovered if only we'd been allowed to look. What does it cost us to prevent new or enlarged areas devoted to human habitation, cattle grazing, or agriculture? What price tag comes with a prohibition on road building and dam construction and all the potentially productive changes new road and dams might indirectly deliver?

These and other lost opportunities are very difficult to estimate with any degree of precision. We can never be certain what we are missing by *not* conducting additional developmental activities in a relatively under-explored region. We might cause ourselves to fail to discover immensely valuable new genetic resources or vast new energy deposits or previously unknown medical marvels...but who knows? It's comparatively feasible to guesstimate how much revenue we lose directly by not putting in a new factory, dam, or farm, but these ancillary and even accidental losses are far more unpredictable. We need to try, in order to make it possible to arrive at any rational decision that incorporates costs and benefits realistically. But this is, quite unavoidably, yet another area where uncertainty fogs up our lenses as we strain to get a clear look at our biggest environmental dilemmas.

With these limitations in mind, I think it's safe to say that we are talking about the most massive, most trans-national, most well-coordinated, and most expensive habitat conservation initiative in history. Would it be effective? Unfortunately, no one can guarantee a positive outcome irrespective of how much money we throw at the problem, because we don't know enough about the myriad aspects of life on this planet to certify that any conservation initiative will work as intended. It's reasonable to presume that if we do all we can to safeguard key habitats, we will save the species that depend on them. But life has a way of surprising us, and not always in a pleasant way...and good intentions don't count for anything. My dating experiences prove this much conclusively. There's no question that, the actual efficacy of our efforts notwithstanding, both the direct costs and the missed opportunity costs would be at very high levels. But "expensive" is a relative term, and we always have to ask: Expensive compared to what? And is it expensive but still worth it?

Despite the fancy prices and significant sacrifices involved in any overarching response to the biodiversity challenge, none of this would likely entail drastic changes in the way most of the world obtains or uses energy, or in the way most of us live our lives on a daily basis. There definitely would be some noteworthy restrictions on jobs and lifestyles for people who actually live and work in or very near the protected areas; but—vast as those numbers are—this would still prove far less onerous than a top-to-bottom restructuring of modern civilized society. Unlike the most probable corrective actions in the climate change area, then, the downside costs of biodiversity-preservation measures *should* be of a lower and more manageable order of magnitude. This ballpark-level distinction could be important when we ultimately get around to making the decisions that I call betting the Earth. It could be a factor, among others, that affects our choice as to what is a prudent gamble and what is not.

So now we have introduced two global environmental dilemmas—climate change and mass extinction—that are in some specific respects very different, yet united by several key features that make them worthy propositions for the highest-stakes gamble in all of history. As we search for a new way of making the most important decisions of all time, we need to be specific about the realities confronting us.

What kind of a world-saving book would this be without a numbered list or two? So let's make a list of some of the characteristics shared in common by both of our Big Bang environmental threats. These unifying features will make it possible for us, in the concluding chapters, to consider both climate change and mass extinction as related but separate issues within the context of our gamble for the Earth's fate. Here then are some of the shared traits of the climate change and mass extinction phenomena that arguably threaten us today:

1) The potential to cause enormous and permanent damage to human beings (their lives, their health, and their economic wellbeing), now and in the future.

2) The potential to inflict massive and persistent harm on Earth's living things in general.

3) A tendency to exert most harmful effects—if at all—very gradually and in a subtle manner, so as to be imperceptible to ordinary human observation.

4) Effects that are not narrowly localized, but instead widely diffused and often primarily affecting people and places far from the United States of America.

5) A greatly prolonged time frame for damage to take place, perhaps extending many decades or even centuries into the future, and perhaps not at all.

6) Some credible scientific evidence both in support of and in opposition to the problem as a major threat.

7) The existence of possible corrective courses of action, which would be very expensive and disruptive to one extent or another, yet not certain to be effective.

With this somewhat disturbing background in place, I will now turn to the work of three of the preeminent thinkers of all time. This portion of the book will briefly examine the crucial insights they contributed to human thought regarding *uncertainty*. I will summon up these three towering figures from our intellectual pantheon to deliver some much-needed help. After all, we're trying to come to terms with all the doubts tangled up with our most profound environmental dilemmas. Who couldn't use a little expert assistance at a time like that?

The next three chapters will, I believe, provide you with some fresh and wonderful ideas on this long-unsolved, yet not unsolvable, puzzle of how to live with uncertainty and danger. In each instance, I will focus on a historic breakthrough in understanding the nature of uncertainty, and the optimal method of dealing with it. The insights from these great trailblazers will point us forward to a way of understanding the role of *incomplete* and *unclear* information in life, and indeed in the entire universe. I am convinced that their legacy can assist all of us in developing the best possible response to the most momentous of all unanswered question marks. Their penetrating vision will enable us to see clearly through all the haze to the core of such vital contemporary issues as climate change and mass extinction. When you're gambling with the future of the whole planet, that kind of advantage can make all the difference in the world.

CHAPTER 4

Pascal and Decisions in the Midst of the Mist

Blaise Pascal (1623-1662) was an amazing person of immense abilities. The enduring contributions from his brief but extraordinary life are numerous and varied.[1] Among other things, Pascal produced great innovations in fields as disparate as hydraulics and geometry. For purposes of this book, we will draw on his breakthroughs in probability theory and decision analysis, with special focus on his concept which has become world famous as Pascal's Wager. But first, let's look for a few moments at his personal story.

Blaise Pascal was born on June 19, 1623 to a family of merchant-class nobility in the city of Clermont-Ferrand, the capital of the Auvergne region of France. As recalled by his sister Gilberte, Pascal was a gifted child with a curious nature, who astounded adult visitors to the Pascal household with profound questions atypical of a child his age. While he was never formally educated at a university, Pascal was taught mathematics and classical languages by his father Etienne Pascal. Etienne caught a glimpse of his son's genius when one day he found that twelve-year-old Blaise had arrived at Euclid's conclusion—that the sum of the angles of a triangle is equal to two right angles—completely unguided. Pascal's insight and intellectual ambition at an early age were signs of great things to come.

By Pascal's sixteenth birthday, he had published his first essay explaining the properties of conic sections as they relate to a circle. After writing *Essai pour les coniques*, Pascal began to attract the attention of scholars in the fields of mathematics and science, and evoked negative reviews from the famous Rene Descartes, then forty-four years old, who was jealous of Pascal's prodigious abilities. When the two great minds of the 17th Century eventually met in 1647, Descartes reacted with an air of condescension, apparently still threatened by Pascal's genius.[2]

A follower of the Catholic faith's fundamentalist Jansenist sect, Pascal struggled with his spirituality throughout the beginning of his adult life. Jansenists followed the teachings of Jansenius, who called for a reshaping of Catholic thought and conduct to more closely mimic that of St. Augustine of Hippo. The Augustinian position held that to achieve a state of grace, people must use their own free will to live as closely to God's ways as possible, because a life of sin can only enslave human beings to be driven by the constant need for self-gratification.[3]

On the 23rd of November, 1654, any doubts as to Pascal's spirituality would go up in flames. On that electrifying evening, Pascal experienced a dramatic spiritual awakening, *la nuit de feu* or "the night of fire," that would change him forever. In his sister Jacqueline's words, he ceased to be defined in terms of mathematics or other worldly pursuits, but instead, became a true Christian believer of devout faith.[4] From that moment until the day he died, Pascal always kept with him a memorial of the night of illumination, his night of fire, on which he renounced his life before God with fiery intensity and embraced a new calling to serve Christ.

Pascal never ceased his intellectual pursuits after his religious awakening, but conjoined religion and mathematics in a way that would have a meaningful and lasting impact on the world. For Pascal, an intellectual factor had to be present in order for him to embrace theological controversy, and the school of Jansenius gave him just that. He wrote "[i]f we submit everything to reason, our religion will be left nothing mysterious or supernatural. If we offend the principles of reason, our religion will be absurd and ridiculous."[5] Pascal injected his religious fervor with mathematics, a union that gave rise to what is arguably his greatest work—the Wager.

The Wager was contrived by Pascal when he was approached by a friend who enjoyed games of chance. His friend, the Chevalier de Mere, was a bon vivant who recruited Pascal to determine the odds in various gambling scenarios. From this somewhat seedy origin came the seed of modern probability theory. Pascal developed the notion that future events could be predicted with mathematical precision using numerical odds—a feat unsuccessfully attempted by both the Greeks and the Romans.[6] Pascal's efforts in probability theory formed the basis for his Wager, the foundation upon which modern risk analysis is built. The Wager was one more piece of evidence illustrating Pascal's brilliant union of mathematics, decision analysis, and religion that sparked to life on his night of fire. When he died at age 39 on August 19, 1662, he was still carrying the memorial of the night of fire next to his heart, sewn into the lining of his coat.

When a killing frost ends any life still in its budding springtime, our tears always water the forever-unripened and unharvested fields of what might have been. But, along with the early farewells of Mozart and Raphael, the hasty loss of Pascal's light cost us more unkept promises than any other.

Pascal did leave behind many legacies, including his Wager, and that's what we need to consider now as it relates to the subject of this book. In its most general application, Pascal's Wager is a framework for formulating a wise decision concerning a momentous matter when there are some major unknowns relevant to the equation.[7] In other words, Pascal's Wager is designed to help us decide on a rational course of action when we have to make very important choices despite imperfect and incomplete knowledge of some key factors. It has been applied to a broad range of problems, but it was originally conceived in a religious context, which I will outline here to illustrate what Pascal had in mind when he created this idea.

I use the word "outline" deliberately, because Pascal's Wager in its original form is one of the most astonishing examples of concise expression in all of human history. The entire idea now known as Pascal's Wager is completely contained within just a single paragraph of his collection of thoughts (note 233 of his *Pensées*), along with other important and related insights! But the *Pensées* weren't published until after Pascal's very untimely death, and he never put them in any particular sequence, nor did he finish writing and polishing all of them. They really were mostly a compendium of short working notes Pascal had jotted down during the last few years of his life while he was busy writing a major work on defenses of Christianity. He may not have even envisioned publishing them. As helpful aids to the actual project he was working on, the notes reflect the many times he returned to the same ideas, inserting additional thoughts, modifying or crossing out others, and generally developing and refining the material for his treatise. The *Pensées* weren't a work in progress so much as they were its intellectual scaffolding and draft plans. After all, that's mostly what "thoughts" are—means to an end rather that an end in and of themselves, and a journey rather than a destination.

The original piece of paper that became note 233 is itself a remarkable document. Pascal's handwritten comments, insertions, and corrections run all around the edges of the page like determined hounds chasing a wily fox through a thicket, and they continue their hot pursuit onto the paper's back as well. My sainted mother, Mae Kunich, often wrote letters in the same way, and with a document that closely resembles an Escher print, it can be a bit of a challenge even to figure out where one sentence ends and another begins. Because Pascal never "finished" note 233, and maybe never even intended to, it contains some odd and seemingly inconsistent words and phrases. It's easy to punch holes in it, mostly because Pascal probably meant it to be just his own private preliminary scratch pad and not a public punching bag. But I think it is safe to say that this single scrap of paper is the greatest and most influential by-product of all time. In this one paragraph, written as a tool to help him create something else, Pascal shot a volley of grappling hooks into the future, with prongs that included the cutting edges of probability theory, decision theory, infinity analysis, game theory, risk management, and other stunningly penetrating points.

Pascal was interested in persuading skeptics that rational, reasonable people should choose to believe in God. He assumed that God in some sense demands that people believe in Him and that He rewards such belief with the infinite and eternal blessings of Heaven in the afterlife. The "wager" in Pascal's original formulation comes from the fact that we don't know with certainty whether God exists, and so we are betting, so to speak, on this in making our choice to believe in Him or not. Pascal easily could have added other related and relevant uncertainties, such as whether God rewards belief, standing alone, with eternal bliss in Heaven or also requires compliance with the commandments, and/or whether God punishes either disbelief or violation of the commandments through eternal torture in Hell. In light of all that has grown from his "wager" it is incredible that it was contained in only one paragraph of Pascal's posthumously collected *Pensées*, and that this solitary paragraph actually includes three related thoughts, of which only the last has come to be known as "Pascal's Wager." The *Pensées* were first published in 1670, and that lone paragraph that became note 233 continues to move people more than a third of a millennium later.

Pascal reasoned that it was a good and prudent "bet" for people of skeptical inclination to choose to believe in God. Here's why. If we "bet" on God's existence (and on God's rewarding those who believe in Him via everlasting and unimaginably immense reward), we will sacrifice virtually nothing except possibly the time and effort necessary to pray, light candles, attend religious devotional meetings, and other manifestations of belief. We'd risk next to nothing by betting on belief in God, with the prospect of infinite reward if our bet is correct and God exists. You don't have to be a math major to see that, given Pascal's assumptions, this would be a very wise bet to make. He summed up his conclusion by writing as follows:

Let's weigh the gain and the loss in wagering that God is. Let us estimate these two chances. If you gain, you gain all; if you lose, you lose nothing. Wager, then, without hesitation that He is.... But there is here an infinity of an infinitely happy life to gain, a chance of gain against a finite number of chances of loss, and what you stake is finite.

Because Pascal's Wager is so briefly set forth in the unfinished working papers that after his death became his *Pensées*, there are some obvious points where his formulation is incomplete or makes needlessly extreme assumptions. During the next few pages, I'm going to take the liberty of adding my own details, alternatives, and (I hope) reasonable variations to the stark framework of Pascal's Wager. I apologize to the purists I am about to offend, and in no way do I intend to imply that I am fit to improve upon Blaise Pascal's inestimable achievement. I'm going to suggest some alterations to the wager, but only because this will make it a closer fit to the global environmental dilemmas at issue in this book. By making these rather clumsy adjustments, I think it will be easier for me to show, in Chapter 7, how a framework similar to Pascal's Wager could help us resolve these very earthly—but momentous—puzzles. So please bear with me as I play God (if you'll pardon the expression) with Pascal's Wager.

Pascal himself didn't address the scenario in which God requires more than belief alone but also adherence to a list of scripturally-rooted commandments; but if he had, the analysis would not have been much different. Under the "obedience to commandments" situation, our compliance would cost us a certain finite amount of enjoyment and gratification regarding life's carnal and physical aspects. The list of temptations in which we might decline to indulge could take a variety of forms depending on how one view's God's commandments. In this variation on Pascal's original wager, we might "bet on God" by refraining from unbridled avarice, gluttony, sexual hedonism, sadism, dishonesty, mendacity, theft, manipulative behavior, and exploitation of others. In the alternative, we could make the opposite bet and behave in whatever manner we prefer, fully yielding to such temptations (at least within the limits of human civil and criminal law), under the assumption that there will be no negative consequences for us after we are dead. Is either of these a prudent bet, and is one preferable to the other, given our uncertainty as to whether God is watching and keeping track for an eventual day of reckoning? It works out, given the givens, in roughly the same way as before.

Pascal knew that such dilemmas are deeply troubling to people who long for certainty wherever and whenever possible, let alone when making life's most important decisions. Pascal constructed his original wager to demonstrate a powerful means of making appropriate allowances for unknowns and unknowables when we must choose "in the dark" between options with very different potential outcomes. He reasoned that wise decision-making in an environment of significant doubt required systematic consideration of the possible positive and negative results from any of the choices available, plus some factoring in of the relevant probabilities. This was the birth of both decision analysis and probability theory, on one bit of scratch paper.

In working through his wager, Pascal made some important threshold assumptions about the situation in which there was no God. Again, essentially the same assumptions would apply if there were no afterlife, or, if God exists and there is an afterlife but there is no dramatically different eternal punishment or reward predicated either on a person's belief in God or his or her lifetime compliance with certain commandments. Under any of these sets of circumstances, what might be the positive outcome from a correct bet? In Pascal's original "belief only" wager, if it turns out that there is no God and we have bet along those lines by choosing not to believe in God, the payoff would be either a little time and effort saved (by not praying or going to church) and/or all of the enjoyment and gratification available from, as Pascal phrased it, "those poisonous pleasures, glory and luxury." Worth having, perhaps, but not infinite.

Likewise, in my "obedience to commandments" variation, a correct bet against a non-existent God could free us to lead a life of unrestrained hedonism. We might opt not to adhere to any particular group of moral or religious rules, but rather to live life to the fullest with utmost enthusiasm for selfish, carnal, materialistic activities. No doubt Pascal would have included such things in his category of "poisonous pleasures." Again, this could be viewed as a small or significant upside, depending on the value we assign to the time diverted from prayer to pulchritude and the gratification gained from wanton pleasure-seeking. However, these transitory plusses could also come bundled with notable disadvantages associated with such behavior (e.g., sexually transmitted diseases, lack of genuine friendships, physically and financially ruinous drug addiction, broken families, shallowness of relationships, etc.). Significant or not, Pascal saw the payoff from a correct bet against a non-existent God as finite, strictly limited in both time and magnitude.

Conversely, Pascal thought that an incorrect bet here would have fairly limited negative consequences and some worthwhile positive advantages. This is where we're talking about a decision to believe in God when, it turns out, God does not in fact exist. Under the original formulation of the wager in which belief alone is involved, it could be that the practice of prayer and meditation would itself produce some real, if limited benefits, such as improved outlook, peace of mind, and serenity of disposition derived from quiet contemplation and devotion to noble thoughts. Pascal specifically listed several noteworthy advantages (from belief in an illusory God). In Pascal's view, our consolation prize would consist of these finite rewards: "You will be faithful, honest, humble, grateful, generous, a sincere friend, truthful."

Similarly, in the "obedience to commandments" variation I've suggested, there can be worthwhile finite advantages from a pious lifestyle, even if the prospect of eternal, infinite bliss proves illusory. For instance, a person leading a moral life of self-control, generosity, sexual restraint, moderation or abstention in the use of drugs or alcohol, and honesty may not enjoy all the earthly delights available to the hedonist, but he or she may actually be happier, healthier, more loved, more secure in long-term interpersonal relationships, and more fulfilled. For Pascal, if the unknowable truth is that there is no everlasting reward or punishment for our belief in God or our adherence to religious rules, a person would probably neither gain nor lose decisively regardless of guessing right or wrong in deciding how to live. It could make a difference, of course, but in most cases there would not be an overwhelming preponderance of evidence favoring advantages or disadvantages based on a decision either to yield to every temptation or to try to live according to a strict moral code.

Pascal contrasted this with the results obtained if there in fact turned out to be a God (and, I would add, an afterlife and a nexus between a person's mortal behavior and post-mortal experience of unending bliss or agony). He assumed truly extreme, unimaginable intensity of eternal reward in Heaven (and we might have added similarly infinite eternal punishment in Hell). In this situation, a correct bet on a lifetime of belief in God (or belief plus obedient rectitude and restraint) would pay enormous and everlasting dividends infinitely far beyond any transitory pleasures a person might deny himself or herself during a brief mortal life. But an incorrect bet here would result in the ultimate "bad bet." The pleasure seeker's indulgence in the ephemeral attractions of a little time saved—or even of a lifetime's maximum load of lust, greed, sloth, gluttony, and dishonesty—would come at the highest possible price: unending eternities of banishment from the infinite rewards of Heaven. Pascal didn't even mention the additional possibility of everlasting, infinite hellish torture and loss, but that bonus feature could only bolster the argument he was making. Pascal saw this vast dichotomy of potential consequences as a crucial factor that must be adequately addressed in making a decision.

Pascal's Wager, in the estimation of Blaise Pascal himself, pointed to an obviously optimal decision in light of the vital, relevant unknowns that could not be definitively resolved. He argued that every action, *including nothing more than inaction*, is in fact a form of decision-making under these circumstances; it is impossible to abstain or to refuse to decide. He asserted that there were far greater benefits and losses attendant to the (unknowable) actual reality if there were a God who assigned eternal reward based on our belief than if there were not. He maintained that his wager made it clear, despite the irresolvable questions, that much more was to be gained and much less to be lost in "betting on God," and much less to be gained and much more to be lost in "betting against God." Pascal, as you will recall, reached this conclusion in these words: "But there is here an infinity of an infinitely happy life to gain, a chance of gain against a finite number of chances of loss, and what you stake is finite."

Thus, Pascal's Wager can be thought of as a way of analyzing the upside and downside potential associated with each of the probable sets of unknowable facts, and using the results to determine which option presents the best combination of risk and reward. In Pascal's formulation, the situation included an infinite value (eternal reward of limitless greatness in Heaven) which had to be weighed against only finite considerations and alternative outcomes. As he recognized, in such a situation, the infinite value "dominates" all else and determines the proper decision. The basic structure of the wager, with appropriate variations and adjustments, can also be employed in settings far removed from the original religious context, as I will show near the conclusion of this book.

Pascal's Wager has been criticized and defended ever since 1670, and there's no end in sight. I'll just take a moment to touch on some of the points of contention.[8] Most (but not all) of the criticisms are aimed at Pascal's assumptions, such as infinite reward for belief in God irrespective of one's motives for believing, or the possibility of infinite utility at all,[9] or whether one can truly decide to believe in God, or whether God might in fact reward those who wisely remain skeptical or agnostic. Critics point out that there could be alternative actions or decisions that would evoke infinite benefits or infinite loss, either in addition to or in lieu of a choice to believe in God. Further, the wager seems to neglect the possibility of many Gods[10] (with conflicting and mutually incompatible, inconsistent systems for assigning rewards and punishments), or of just one God, but a different one from the God of Christianity, who punishes believers in the Christian God. Pascal also ignores the prospect of an all-merciful God who grants infinite blessings to everyone, irrespective of their disbelief, or of a partisan God who imposes different rules on, and grants different rewards to, His "chosen people" (whoever among many candidate groups that might be). And some critics attack Pascal's insistence that we assign a probability greater than zero to God's existence.

For God's sake, that's enough of that! Let's come back to Earth. This is not a book about the age-old question of divinity, thank God, and so it will suffice for my purposes to get the basic structure of Pascal's Wager set up and available for us to tinker with here on the ground. Pascal gave us a powerful mechanism for analyzing important "decisions under uncertainty" quite apart from anything dealing directly with theology. It is this broader application of Pascal's seminal insights that we'll use for the remainder of this book. I'll leave "Nearer, my God, to Thee" to the hymnals and instead steer straight for our home turf right here on Planet Earth. That way, I won't have to referee the debate between the defenders and debunkers of his wager in its archetypal form. Whew! Plus, I plan to wind up with an analytical engine that will help us decide, rationally and logically, the biggest dilemmas threatening our global environment.

Let's bring Pascal's Wager a lot closer to home (maybe a bit too close for comfort for some of us) with a real-life example from the personal experience of the humble, somewhat neurotic, author of this book. Maybe you can relate. Every year, as my birthday approaches, my doctor sends me a reminder that I'm due for my annual physical check-up. Part of this recurring ritual features a dialogue with myself that consists of an internal debate over whether it's worth it. This little silent argument nicely follows Pascal's analytical framework on a personal, individual level.

There are some knowns and unknowns involved in my, or our, problem of whether to get a yearly physical. Among the knowns are the following minus factors: It will take about two hours' worth of time from our schedule to go to the doctor's office; it will cost us a small amount of money; and we don't enjoy the exam itself (particularly the parts consisting of the doctor's rubber-glove-covered fingers and one of our orifices). There are also some unknowns, both plus and minus: The exact amount of time the exam will require; precisely how uncomfortable the process will be; how freezing-cold or just chilly-cold the doctor's hands will feel; and how annoying or pleasant we will find the time we spend in the office and in the examination room waiting for our turn.

But there are more unknowns that are far more significant: Will the exam reveal a serious, even life-threatening medical condition? If it does, will the doctor discover our condition early enough to initiate a course of treatment that will save our life and avoid a prolonged period of intense pain before an untimely death? Or will the discovery of our problem only bring us a time of additional worry, discomfort, and expense as we hopelessly flail against what turns out to be an incurable and terminal affliction? In other words, among other alternatives there is some possibility that our choice to have the physical will be rewarded with many years, or even many decades, of additional happy life and good health, and spare us a premature and wretchedly agonizing death...both of which also have major consequences for our loved ones and anyone else who depends on us for material or intangible support.

Pascal teaches us that we should weigh both the size and the likelihood of each factor before we make our decision. When we look at the situation in this light, it becomes clear that any minor inconvenience, expense, discomfort, and annoyance it will (almost certainly) cost us to undergo the physical exam is finite and quite small. All of this is overwhelmingly dwarfed, or dominated, by the possibility that the exam could save us from a premature and excruciating death while granting us a much longer, more enjoyable and productive lifespan. Even if, as is thankfully the case, it is relatively unlikely that any given exam will reveal a dreadful (but treatable) hidden medical problem waiting to kill us, the sheer incalculable magnitude of the benefits we'd receive from this early intervention makes it a wise choice to have the physical. A small chance that the exam will spare us a horrible and early death is good enough for us to choose to accept the minor costs and inconveniences of the exam.

Even a low probability that an annual physical will give us another 20 or 30 good, healthy, fruitful years to enjoy with our loved ones and fend off unbearable pain and suffering is sufficient—under the helpful lens given to us by Pascal—to outweigh any combination of comparatively insignificant countervailing factors on the equation's other side. Don't you agree? After thinking about our options in these terms, we take a deep breath and tell ourselves, "Better safe than sorry," and make the darn appointment, often without even realizing that we have good old Blaise Pascal to thank for our common sense.

In this way, Pascal's Wager allows us to conceptualize the benefits and risks inherent in any determination so we can appreciate whether certain "bets" are wise or foolish. In essence, Pascal's Wager is a guide to rational gambling where the stakes are extraordinarily high, and where the decision-makers must find a reasonable way of dealing with huge unknown factors. The decision-maker can consider the probability and magnitude of the outcomes from each combination of extreme values for each variable in comparison with one another to evaluate whether there is a greater risk from non-action or a greater reward, and vice versa.

It's humbling for me to admit that an idea as seminal and enduring as Pascal's Wager originated as little more than an aside, a brief intermediate working note, in the prodigious output of Blaise Pascal. It's true, though. This fascinating and revelatory thought proved just one small piece of the work of the man whose intellectual sparks ignited probability theory, decision theory, game theory, infinity analysis, uncertainty analysis, risk management, and practical contributions from the calculator to hydraulics. To say that Pascal's work had legs is like acknowledging that a millipede has a leg to stand on. His brainchildren and their innumerable descendants continue to this day to shake the world. But what specific relevance does Pascal offer to the problem at the heart of this global gamble we're venturing right now? Before we can answer that I need to mention some of the wager's modern outgrowths that can also be useful in untangling our world's environmental knots.

Pascal's pioneering forays into decision-making in the midst of uncertainty have several related but somewhat divergent branches. We can reach for them as we try to swim out of these swirling rapids. In this book, I am focusing primarily on probability theory as a modified form of Pascal's Wager because I think it's a good conceptual fit for the nature of the environmental puzzles at hand. All things considered, I believe we would be justified in applying decision theory to the question of how to respond to potentially devastating, yet uncertain, environmental threats. However, if we prefer an analytical framework less dependent on probability, we could instead use such tools as Info-gap and Minimax. Allow me a few moments to introduce both of these alternatives, very briefly, just to open those doors to anyone who cares to see where they might lead.

I think it's fair to say that decision theory is a direct descendant of Pascal's work on probability and choice-making, including Pascal's Wager. Simply put, decision theory is an analytical framework for determining in a rational, logical manner the best choice in any situation in light of the relevant risks and rewards, the likelihoods of various occurrences, and incomplete or uncertain information. Decision theory of this type, i.e., the type aimed at developing the optimal course of action under a given set of circumstances, is called "normative" or "prescriptive" (as opposed to the related variety, which is "positive" or "descriptive") because it is focused on what we *should* do ideally and not on what ordinary people *actually* do in their usual practice. This distinction recognizes the realistic point that many people do not arrive at decisions (even the most important ones) logically, objectively, rationally, and systematically. I'm very grateful that I don't fall into that unfortunate category…much.

The specific form of decision theory reflected in Pascal's Wager is called a "choice under uncertainty." Where one or more of the variables relevant to a particular decision can't be precisely determined, Pascal and his heirs among the decision theorists address that uncertainty through use of probabilities, including the notion of "Expected Value" (or EV in mathematical jargon). Expected Value might sound complicated, but in essence it's just what the name implies; in light of the probability of something happening, and the value or size that something would have if it does happen, EV is the one number that best describes the value we would expect it to have, on average, in the long run. We can calculate the EV of any potential event simply enough, just by taking all of the various *possible* values the event would have (both positive and negative) if it does in fact happen, and multiplying, one by one, each of those values by the fraction or decimal number reflecting the corresponding probability of every specific value actually taking place. Then we'd add them all together to arrive at the Expected Value for the entire event.

For example, what if I come into a game of chance with no money, but I'm offered a chance to play? I can toss a one-dollar coin in the air with the proviso that I get to keep it if it lands heads-up, but I lose the coin if it lands tails-up. What is the EV of choosing to play this game? The EV of this coin flip gamble is zero, that is, no change in my initial situation. Here's why. The winning or positive outcomes would boil down to +$1.00 (the result wherein I win one dollar every time I throw a heads outcome) multiplied by 0.50 (because the coin has a 50 percent probability of coming up heads), or a partial (positive) value of +50 cents. The flip side of this (sorry) is that the negative or losing results amount to -$1.00 (the loss of one dollar for every tails I throw) times the same 0.50 (because a fair coin also has a 50 percent chance of coming up tails), or a partial (negative) value of -50 cents. The overall EV is then the sum of all (both) of these partial EVs, of +50 cents plus -50 cents, or 0.00. This big fat zero of an overall EV tells me that if start with nothing and I flip a coin numerous times, the net overall result of all those one-dollar wins and losses, in the very long run, will be exactly nothing in my hand. Big deal, huh?

Let's make this more realistic and look at the EV of a dollar wager we might place on a single spin of a standard American roulette wheel. The wheel has 38 equal-size slots into which a ball might drop, with slots numbered from 1 through 36, plus 0 and 00. A bet on any single one of these numbers loses the entire value of that bet if any number other than that one number winds up with the ball. But if the ball lands in the slot on which we've bet, we win an additional 35 times the size of our wager, and we get to keep our original wager. So the probability of any one number receiving the ball is 1 out of 38, and the probability of any number other than our one specific number winning is 37 out of 38. The winning outcome of a one-dollar bet would then have a value of +35 (the size of the winning payout) multiplied by its probability of 1/38, or +0.921. The losing outcome of that very same one-dollar wager has a value of -1 (the amount we lose if our number we've bet on doesn't come up) multiplied by the probability of losing (37/38), or -0.97368. These are the only two possible outcomes in a roulette wager—either the ball lands on the number we've bet on, or it doesn't—so we don't need to factor in any other contingencies.

The overall EV of any bet on the roulette wheel is just the sum of these two (and only two) possible values, or 0.921-0.97368, which equals an Expected Value from our bet of -0.526, or, expressed instead as a percentage of our wager, -5.26 percent. This negative number means that for every dollar we bet on a roulette wheel, on average if we place the same wager a great many times and keep track of all the individual wins and losses, we should expect to lose a little more than a nickel. Or, if you prefer, for every $100 we bet over a very large sample size of thousands of separate spins of the wheel, we expect to lose $5.26, on average. That's the sad truth about what a negative EV signifies. The wager is, on balance and in the long run, a losing proposition. Our negative EV certainly doesn't mean we will lose on any particular spin, or even that we will lose money over several entire days or weeks of gambling, but eventually the odds are that we will lose roughly in accordance with that figure, -5.26 percent of all we wager. Hey, someone has to pay for the artificial volcano outside the casino.

Decision theory uses EV as a way of differentiating among a range of options, each of which has more than one possible outcome, and each of which has its own probability of happening. All those potential results and associated likelihoods, whether large or small, can be reduced to a mathematical problem that squeezes much if not all of the subjectivity and guesswork out of the equation. According to some proponents of decision theory, the rational choice is to select the alternative with the highest overall EV among all available options. Under this view, in the long haul and on average, the highest EV choice will outperform all the others, and thus the optimal decision would be to go with it. Decision theory, oversimplified in this way, is a method of playing the odds...like Pascal's Wager itself.

Decision theory can also be applied to closely related situations known as "decisions under risk," in a practice often called risk management. Risk management is certainly an appropriate way of conceptualizing the global environmental threats we're wrestling with in this book, because it's hard to imagine a greater risk we need to manage. I'll return to this in Chapter 7, where I'll take a closer look at the specific questions wrapped up in any decision under risk. Once you chip away all the camouflage, risk management boils down (or boils over?) to an examination of *what* can possibly go wrong under a given set of circumstances or decisions, the *probability* of each possible bad outcome actually happening, and the *size* of the negative consequences associated with each outcome in the event any of them do materialize. It's a matter of listing the possible harms, and then assigning both a likelihood and a magnitude to each one. Once we do this, we should have some idea what level of risks we're dealing with, and theoretically we ought to be able to make rational decisions that logically and appropriately take those risks into account regarding the optimal course of action. Of course, uncertainty enters into the risk management equation, because if our information is inaccurate or incomplete in significant respects, the probability that we will make an incorrect (suboptimal) decision increases. This makes it important that we incorporate uncertainty into our risk management decision-making in a way that allows us to absorb the impact of unknowns.

Pascal anticipated the full development of risk management/decision theory, the meaning of infinity, and probability theory in the way he structured his wager. His wager can be represented by a decision matrix in which there are two possible states of reality (God exists, or God doesn't exist) and two alternative decisions available to us (believe in God, or don't believe in God). Pascal posited that the Expected Values of the outcomes that follow from three of the four possible combinations of these factors are finite. Only one of the four potential results has an Expected Value that is infinite—the consequence of believing in God when God in fact does exist. Under decision theory, this lone infinite EV outcome "dominates" all the finite EV possibilities and points to the optimal decision, and the only rational decision given the premises.

Decision theory has its critics, as does Pascal's Wager, and its alternatives. Some commentators have pointed out that decision theory may be poorly suited to identifying and appropriately incorporating the influence of unexpected variables or "unknown unknowns." That is, we may not be able to anticipate all the relevant factors and we may not have an accurate estimate of all the relevant probabilities. Despite our best efforts to identify, assess, and include all the pertinent factors and to build in room for the unknowns we're aware of (the "known unknowns"), there's always a chance that we won't capture everything. There could be new and significant variables that pop up, consisting of either pleasant or (all too often, in my experience) unpleasant surprises, and our EV calculation may miss such things, to the detriment of the decision it identifies as optimal. Like any model, a decision theory will have its flaws, and the prospect of unforeseen events arising "outside model" can be an important one.

This is an important idea with special relevance to the environmental context, so it's fitting that it has a naturalistic name—"the Black Swan theory."[11] The name stems from the situation hundreds of years ago in which Europeans were unaware that swans could be any color but white. People from that time and place who were interested in swans might have postulated models to predict various characteristics of swans, such as their weight, length of neck, wingspan, longevity, etc., but they probably would not have included the possibility of black swans in their models. Black swans weren't on the radar scope, and no one within the circle of investigators had ever seen or heard of one, so the probability of ever encountering such an unprecedented situation would have been very easy to miss. Until the 18th Century, when Europeans discovered black swans in Western Australia, they would have been considered as impossible and unworthy of incorporation into any model as unicorns and winged horses. They would have been considered outliers (if anyone even thought of them at all), deemed to be impossible and thus not part of any realistic set of assumptions built into a model for making decisions.

Black Swan theory in general holds that decision theory is vulnerable to omitting or minimizing the potential impact of unanticipated variables—the unknowns we don't know are out there. Sometimes unprecedented and unforeseen situations jump out at us and (almost by definition) catch us unawares. Under Black Swan theory, these "unknown unknowns" can go far beyond the expected variations we've anticipated and incorporated into our decision model, and can have a profound effect on what decision is optimal. Black swans are game changers. And believe me (as a lifelong Chicago Cubs fan), when a black swan (or even a black cat) blindsides you, the game doesn't change in your favor.

When decision theory focuses on a limited range of possibilities and fails to make room for the potentially huge impact from outliers that are assumed not to exist, there is a risk that we might fall victim to the "ludic fallacy" where we mistake our model for the reality we want it to reflect. The ludic fallacy is always a danger, because it's so easy for us to forget that we can't ever know *all* the relevant information or accurately incorporate *every* source of data over all time and all circumstances. The problem may not be apparent for a very, very long time. Under anything approaching ordinary circumstances, our decisional model might be quite adequate, and all goes smoothly. It's so tempting for us to grow very pleased with our own brilliance and our clever model, and we might wear ourselves out patting ourselves on the back. But what happens when that funny-looking "ugly duckling" one day shows up as a black swan and demolishes our assumptions and everything we've built on their faulty foundation?

Black swan events are usually conceptualized as very rare or unprecedented, difficult to predict, outside the range of our normal expectations, and of vast impact when they do materialize. Unless our decision theory is designed to handle those extraordinary shocks when the "impossible" becomes reality, we can find ourselves unprepared to deal with a perilous crisis. Some analysts have tried to improve on decision theory to develop ways of stopping those black swans and their unknowable distant cousins from catching us asleep at the paradigm.

One competing framework that has been developed as an alternative to decision theory is Info-gap analysis.[12] Here's the core idea behind Info-gap analysis: It is a method of making decisions under conditions of uncertainty, sometimes called "severe uncertainty," in a way not based on probabilities. Because of gaps in the information we have available, we may not be capable of fully understanding the systems we're working with, and we may fail to give proper attention to surprises of various types (such as black swans). Info-gap analysis is designed to deal with this problem. Here's a very short summary of how it's structured.

Info-gap analysis begins with a non-probabilistic quantification of uncertainty. Info-gap treats uncertainty as a *gap in information* (which can be managed) rather than a probability. The "horizon of uncertainty" measures the distance between a point estimate of a given parameter and the universe of possible values for that parameter. Next, we develop a model of the system in question based on our knowledge and on any uncertain elements, as well as the decision to be made. Finally, we come up with a set of performance requirements, including acceptable values we either require or want to achieve. The performance requirements might include the concepts of "satisficing," or doing well enough to meet certain critical requirements, and sometimes "windfalling" requirements for handling better-than-expected outcomes. The uncertainty model, the system model, and the performance requirements are ultimately combined to develop two decision functions that lead to a choice of which course of action to follow—the robustness function and the opportuneness function.

In general, the robustness function points at the largest horizon of uncertainty we can tolerate under the circumstances, i.e., a margin for error measurement. The concept centers around the maximum extent to which we can be wrong in our data, our modeling, and our actions and yet still arrive at an acceptable result—a "satisficing" performance requirement. Here's the rule of thumb for the robustness function: When operating in an environment of severe uncertainty, we should prefer decisions that yield an acceptable outcome over a wide range of errors and uncertainties as opposed to an alternative that only produces a satisfactory result within a very narrow margin of error. In other words, when in (lots of) doubt, go with the alternative that keeps you safer in the broadest array of possible contingencies.

On the other hand, the opportuneness function looks to the smallest horizon of uncertainty we can have and still leave open the possible opportunity for us to reap the benefits of "windfall" good-fortune outcomes. In other words, how much error must we allow to make it possible for us to get lucky in a major way with results that are much better than we'd anticipated? This function tries to give a decision-maker the best chance to get lucky and hit the windfall jackpot while still remaining adequately safe and avoiding unacceptable risks. Opportuneness seeks to accomplish this by preferring decisions that tolerate the slightest possible errors as opposed to decisions that demand large deviations in order to make the same windfall available.

Let me make one more point about this concept of robustness, which is a term that comes from game theory, a form of decision theory. It isn't confined to Info-gap analysis. Robustness is essentially a characteristic of a system that measures its capacity to deal with variations, stresses, threats, and other changes in conditions. A robust system is able to absorb changes in its circumstances or variation from its assumptions, including unpredictable "Black Swan" surprises, without suffering unacceptable levels of damage or loss of function. Under decision theory or other alternatives like Info-gap, we can pursue the goal of robust optimization, a goal very much in line with the environmental threats at issue in this book. Wherever possible, we should try to make our plan or our course of action sufficiently robust to withstand an expansive spectrum of altered circumstances, errors in our forecasts, mistaken assumptions, and Murphy's Law moments in general. Robustness is, in a way, expecting the unexpected.

I'll very briefly mention one other non-probability-based decision theory, somewhat similar to Info-gap analysis, and that is Minimax. This system was originally developed for zero-sum games (where one party's loss is literally and exactly another party's gain) within game theory. The goal of Minimax is to minimize the maximum possible loss, i.e., to take the worst-case scenario and improve it to the greatest possible extent. In Minimax (or Maximin, its close relative) we look at the worst-case scenario under each alternative we have available to us, then pick the decision with the worst-case scenario that is *least bad*. Put differently, we rack and stack all of the worst-case scenarios associated with our menu of possible decisions, then opt for the decision with the smallest possible maximum downside, the one that would hurt the least if things go from bad to worst. We grab the option with the strongest Achilles' heel and hope no one shoots an arrow into it. We might think of it as a form of damage control, or making the best of a bad situation. In the environmental context, we assume Nature is our adversary in a two-party zero-sum game and is playing against us, so the Minimax method is intended to lead us to the least-bad worst-case scenario decisional option as we grapple with Nature (for Nature's own good, of course).

Minimax doesn't use EV, but instead employs scenario analysis of key features of each of the various outcomes, including particular attention to vulnerability and what can happen under adverse conditions. This system doesn't look to how much better or worse the possible outcomes might be, just a rank-ordering of which are better than others. As long as we select the "best" worst-case scenario, there is no consideration of the margin of difference among alternatives.

We will be standing on Blaise Pascal's shoulders for the remainder of this book. The wager we're placing on this planet's fate is a direct descendant of his own pivotal gamble. My analysis of the risks, probabilities, and competing decisions owes more to Pascal's brilliant breakthroughs than words can say. But I am a greedy idea-grabber, so I'm not content to piggyback off just one world-class innovator. Why stop at borrowing from one genius, after all, when you can have two or three? So let's continue with the next two chapters to see whether Oscar Wilde was correct when he quipped, "All my best thoughts were stolen by the ancients." I suppose it gives away the answer when I have to resort to quoting yet another brilliant thinker from the past even when merely framing the issue. Well, as the song goes, "I ain't too proud to beg." Oops, I did it again.

CHAPTER 5

Looking for the Missing Pieces of Gödel's Puzzle

Kurt Gödel (1906-1978) towers above most others as one of the greatest thinkers and innovators in the entire history of mathematics.[1] He was born on April 28, 1906 in the Czech Republic in the city of Brno, then part of the Austro-Hungarian Empire. Gödel was raised in a German-speaking household by a Catholic father and Protestant mother who urged Kurt and his siblings to be freethinkers. Although Gödel remained agnostic during most of his early life, he was a self-proclaimed theist, and stated in a letter to his mother in 1961 that "there is much more that is rational in religion than is generally believed, notwithstanding that from earliest youth we are led to the contrary view through books, experience, and bad religious instruction in the schools."[2] A sickly child who constantly clung to his mother, his dependent behavior would carry through into his adult life. Many years later, when attacked by a street gang of Nazi youths in Austria, Gödel was saved by his wife, Adele, who replaced his mother in her caretaking duties.[3]

While working on his doctoral dissertation at the University of Vienna, Gödel became an Austrian citizen, and when Austria was annexed by Nazi Germany in 1938, he was viewed as a German national. Although he was not Jewish and held no political affiliation, he kept company with both groups and found it becoming increasingly difficult to find engagements at which to lecture. Foreseeing a troubling future for Europe in the midst of the Nazi uprising, he decided to emigrate to the United States and accepted a faculty position at Princeton University.[4]

At Princeton, while his wife grew restless in the elite college town and yearned for the bustling cities of the old country, Gödel, a recluse, kept to himself and enjoyed the isolation from family and friends in Europe. Serendipitously, it was at Princeton where in 1942 Gödel would meet a fellow European colleague also escaping Nazi Germany, a man by the name of Albert Einstein.[5]

Brought together by the Princeton Institute for Advanced Study, Gödel and Einstein were recruited to become part of a select group whose sole duty consisted of being required to think. Think about it! The two expatriates had no problem embracing their assigned tasks. They would become two of the most prolific names in physics and mathematics, whose impact on modern science would reshape the way we view the world. Gödel and Einstein became fast friends, yet there could never be a more odd couple throughout all of history. Their respective personalities contrasted in an almost comical juxtaposition that often worked to their benefit. Just as Einstein urged Gödel to revisit physics, Gödel encouraged Einstein not to abandon mathematics. But Einstein had a strong hesitancy regarding mathematics, bemoaning the many ways of making serious mistakes whenever we start calculating. Even today, millions of students in math classes the world over would readily agree, even if they've never heard of Kurt Gödel, or of Albert Einstein for that matter (apart, of course, from the bobble-head Einstein dolls featured in the popular 2009 film *Night at the Museum: Battle of the Smithsonian*).

Einstein proved to be very helpful in a decidedly non-mathematical and non-relativistic way at Gödel's citizenship hearing in Trenton, New Jersey on December 5, 1947. Gödel had studied the

U.S. Constitution thoroughly and was convinced that he'd brilliantly found an apparent loophole that could allow for a dictatorship. Concerned that declaring the validity of this discovery, however logical, would impede Gödel's chances of becoming a citizen, Einstein urged his companion to remain silent on such topics. Unfortunately, the topic could not be avoided during the judge's dialogue with Gödel at the hearing, and Einstein had to intervene in the midst of Gödel's impassioned argument. It probably helped that Einstein was willing to use his celebrity status, and he later obliged by signing an autograph for the judge, to save his awkward and opinionated friend from near-certain deportation.[6]

Gödel's nature was nothing like his friend Einstein's, who was known to be a jovial sybarite. A hypochondriac, Gödel avoided contact with others at all costs and had a deep distrust of doctors and people in general, brought about by a childhood of infirmity. Gödel was gaunt and sinewy due to self-imposed malnutrition. His emaciated figure looked haunting as he peered from dark-rimmed eyeglasses with sunken eyes, wearing his typical attire of an overcoat and scarf, which he would don even in the warm summer months.[7]

It was because of his long-suffering battle with mental and physical afflictions that Gödel would eventually die of malnutrition on January 14, 1978, weighing a mere 65 pounds at the end. His beloved wife Adele had always tasted all of his food for him, at his insistence, to certify that it was safe before he would accept any nourishment. When she at last became too old and infirm to continue to perform her odd duties for him at the dining room table any longer, he wouldn't eat. After all, without Adele there, how could he know his food was not poisoned? Kurt refused to eat ever again, even unto death, starving himself for his inner demons and for his love.

Never satisfied and certainly never retired, Gödel had spent his last years attempting to find new axioms that would settle the continuum hypothesis, a task still unfulfilled to this day. Although he also authored various writings setting forth an argument for the existence of a divine being, he was reluctant to publish these works in order to avoid being hailed as a true believer in a sea of philosophers who favored reason and logic over the mysticism of religion. He feared that his publication of these works would cause his colleagues to discredit his previous works if he portrayed, even quietly, a belief in God. Gödel's ontological work remained undiscovered until after his death, a true loss for the scientific and philosophic communities. Writes logician Solomon Feferman, "[O]ne may wonder how logic might have been different had Gödel been bolder in bringing his philosophical views into play in relation to his logical work."[8] The world will never know, because, to use Gödel's very own words when he described the turmoil he felt over the death of Einstein, "[W]e live in a world in which ninety percent of all beautiful things are destroyed in the bud."[9]

Gödel made several stunning breakthroughs in logic and mathematics, but his most famous and sublime creations were his two incompleteness theorems, unveiled in 1931.[10] Unfortunately, Gödel's incompleteness theorems do not lend themselves as readily to popular understanding as Pascal's Wager. Any attempt to force the theorems into a widely accessible form is fraught with problems. It is very easy to oversimplify, inject errors, mix incompatible principles, and otherwise corrupt the core meaning of Gödel's insights. However, my thesis in this book is that there are some valid parallels we can draw between Gödel's work and the intractable difficulties that confront us in the uncertain world of international environmental law. There may not be a direct, one-to-one correspondence between the context within which Gödel conceived his theorems and the international legal situation, but I think we can legitimately make some useful connections here. Toward that end, I will try to sketch a rudimentary picture of the incompleteness theorems without sacrificing an undue amount of rigor.[11] If I fail, my only hope is that not too many of my readers will catch me in the act.

Let me issue some disclaimers at the outset. No, I don't mean that actual mileage may vary or that objects in the mirror may be closer than they appear. I have to tell you that Gödel's work involves genius-level insights into some of the most fundamentally central aspects of mathematics. If you're at all like me, your comfort level when it comes to math is midway between panic and surrender, so I understand if you're ill at ease with an egghead eye's view of mathematical theory. Formally, I took and received decent grades in college math courses all the way through the various aspects of calculus, but you wouldn't know it to look at my perpetually unbalanced checkbook or my pathetic attempts to make change for a dollar. But I've managed to man-up enough to confront at least the big picture overview, *Classics Illustrated* version of Gödel's contributions. I won't—because I can't—go any deeper than that. So relax and try to enjoy a fleeting drive-by glimpse of some amazing ideas.

Gödel's first incompleteness theorem, in essence, states that every adequate, consistent, and axiomatic mathematical theory is inevitably incomplete. In other words, some true statements cannot be proven true by relying on the axioms and rules within the theory itself. Every consistent set of axioms is essentially incomplete, inasmuch as true mathematical statements can't be derived from the set; they are undecidable, but true, propositions. The classic example is a sentence such as "This sentence is not provable," or "The system will never state that this sentence is not true." Gödel showed that such sentences (now called Gödel sentences) are true, but that their truth can't be proved from within the system in which they are made.

Gödel's first incompleteness theorem can be phrased another way. Inside the boundaries of a context, proof has its limitations. Truth is a stronger, more powerful principle than provability. Some things are not provable as true or false using the rules and axioms within a given system or frame of reference, but are nonetheless true. But we can only know that they are true by going outside that system and drawing on the foreign resources of another world, a different realm of reality.

The second incompleteness theorem makes a related but separate point. It holds that the consistency of any consistent, axiomatic theory can't be proved inside the system. That is, every consistent theory is necessarily incomplete. You can't prove the consistency of a consistent theory without resorting to resources from beyond that theory. Every system of formal number theory that purports to be consistent is necessarily also incomplete.

Let me define some of these terms for those among you who, like me, can't tell your axiom from a hole in the ground. Axioms are commonly taken to be indisputably true principles or statements, but for the purposes of mathematics, axioms are just basic propositions used as a point from which to begin our analysis of everything else. An axiomatic (or axiomatizable) system will typically contain some undefined terms, a number of axioms that make reference to those terms and describe some of their properties, and a rule or set of rules for coming up with new propositions from already existing propositions within the system. Mathematically, we consider an axiomatic system to be consistent if, given the system's axioms and its derivation rules, we can never derive two contradictory propositions. Or, to turn the sublime into the ridiculous, a consistent system won't let you have your cake and eat it too. That's why I consistently hate consistent systems and have an ax to grind against axioms.

I want to emphasize one crucial point about something the incompleteness theorems do *not* claim. Gödel's famous theorems do *not* maintain that it is impossible to prove anything at all from within a system's own limitations. The theorems do not argue for the universality of unprovability. They are not a sweeping denial of certainty in all situations and under all circumstances, even within the limited realm of consistent, axiomatic mathematical systems to which they strictly apply. They have a more restricted focal point. Gödel's famous idea (the first incompleteness theorem, to be precise) is that within any consistent axiomatizable system there always will be *some* statements, or principles, or postulates, that are true but that can't be proved true solely through reliance on the resources available within that particular system itself. But Gödel did *not* maintain that the existence of "some" true-but-internally-unprovable points implies that *all* points—everything and anything—within the system are in the same leaky boat. To put it simply, the fact that you can't prove everything doesn't mean that you can't prove anything.

This idea—that there are some things that are true that we can't prove to be true—applies regardless of how much effort we pour into gathering all the evidence and weighing the relative merits of all relevant factors that might tend to prove or disprove the matter in question. It isn't a matter of difficulty or expense or inconvenience; we're dealing with absolute impossibility. If we're working within a system that is consistent and incomplete, we may be able to prove the validity of many true statements...but not all of them. Irrespective of the degree of our commitment to learning the truth, or how much money, work, and time we devote to this quest, there will always be at least some nuggets of truth that elude us. Of course, if we fail to be diligent, fair, objective, thorough, and persistent in gathering and assessing all pertinent evidence, we will only make matters worse. Poor or biased fact-finding on our part will add to the list of true statements we can't prove...but we can't blame those extra big-ones-that-got-away on Gödel. Our laziness, carelessness, dishonesty, shortsightedness, lack of preparation, and prejudice may well render us incapable of proving the truth of a whole bunch of true principles, but that's not the incompleteness theorems at work. It's just the natural consequence of our own failings. Just ask my mother and she'll tell you all the stories along those lines about me you'd ever care to hear.

To a layperson, Gödel's two incompleteness theorems may not seem to warrant much attention at all, let alone a tumultuous reaction.

But within the realm of mathematics, his work hit with the force of twin nuclear bombs. Even if the relevance of the incompleteness theorems is entirely limited to mathematical theory, they nonetheless constitute, taken together, one of the most astounding breakthroughs ever. They demolished the theories and beliefs of some of the greatest mathematical minds in history, upset many centuries' worth of assumptions and givens, and irreversibly transformed the way in which experts approach the study of numbers. Gödel exposed the fallacy in the notion that arithmetic or other axiomatic mathematical systems could be entirely certain, complete, and consistent. Despite the aura of definiteness, wholeness, and self-contained logical consistency in such systems, Gödel showed that any effort to prove them both complete and consistent was futile. For the first time, scholars were accepting the fact that there could never be an algorithm capable of establishing the truth of *all* arithmetical principles, and that there were *some* truths about numbers that were literally impossible to prove formally.

This was a profound insight that shocked many eminent mathematicians. Even if this were all that could properly be gleaned from Gödel's work, it would still rank with the crowning discoveries of all time. However, Gödel has also influenced theory in fields very different from mathematics. The application of his incompleteness theorems beyond the precisely defined field within which they originated is very controversial and is viewed by some commentators as an erroneous and dangerous misappropriation. I will not pretend to be able to resolve this debate, but will quickly mention some potential broader applications of Gödel's ideas that have gained favor within some groups of theorists.

Gödel's incompleteness theorems have been interpreted or extrapolated by some to suggest fundamental limits to the concept of truth or knowledge, at least within the fields of science and mathematics. To put it another way, no set of axioms can make it possible to deduce all true natural/scientific phenomena. Others have used Gödel's findings to argue that no artificial intelligence or computer can ever be the true equal of human intelligence, because humans are able to unearth new and unexpected truths whereas a computer's knowledge is hemmed in by a rigid set of axioms with which it has been equipped. And some philosophically inclined individuals have derived the notion that we can never truly understand ourselves, because every closed system such as the human brain can only be certain of its knowledge about itself by, in fact, relying on what it knows about itself. That is, rational thought is incapable of reaching the level of final, ultimate truth, including the truth about ourselves.

This avenue, or side street, of Gödel's thoughts can have some disturbing aspects. For example, some argue that Gödel shows us we can't even really figure out whether we are sane or insane. Just as we can't see our faces directly with our own eyes, we can't definitively and independently assess our own sanity. We need to rely on assistance from beyond our own system to do either one. For me personally, this is just one of the myriad important tasks for which I rely on my daughters. Whenever I want to know whether I'm insane, I simply ask Christie and Julie-Kate and they tell me. I'm not always thrilled with their answer, but what can you do?

Maybe in some strange sense we're all basically little kids sitting in the back of a car on a long drive to a vacation spot we've never visited before. How can we know when we've arrived? Not from any resources strictly within the confines of our own knowledge. Are we there yet? Are we there yet? Are we there yet? I guess we can't be sure unless we stop wondering and ask for information or directions from someone who knows more about the area than we do.

Sorry. I'll admit that Gödel's ideas can be troubling on several levels. For many people, including some of the leading mathematicians of his day, it is uncomfortable to accept that something as apparently cut-and-dried as math is actually uncut-and-juicy. For a neat, orderly system of numbers, well-stocked with rules, methods, and nice, solid, known truths, how is it possible that such an elegant system could contain true principles that can't be proved true through skillful, intelligent use of all the many tools the system supplies? It just doesn't seem fair. And it can be very vexing for the purists when pop culture hijacks something as sublime and brilliant as Gödel's theorems and forces them at gunpoint to land on muddy makeshift runways so impossibly far from Gödel's home turf that he wouldn't even recognize the terrain.

Do Gödel's theorems, created as they were for a very specific set of situations (that is, consistent and axiomatic mathematical systems or theories) truly have anything at all to say about other forms of knowledge, or about nature, or about life itself? And how do we ever really know that a specific statement is truly true (yet unprovable from within the confines of a particular theory) when that statement unavoidably eludes definitive proof under the rules by which it is supposed to play? Who are we to decide, in other words, when we are in the presence of a Gödel sentence that is both true and beyond the power of its home theory to prove? I think the key to understanding this somewhat paradoxical notion can be found in another paradoxical notion. And this comes at you just when you thought it was safe to continue reading this book.

Let's think about Gödel's work as a variation on the theme at the heart of a brief but fascinating little novel from the late 19th Century. I want to look at Gödel's theorems as a fantasia that might have been inspired by the story *Flatland: A Romance of Many Dimensions* by Edwin A. Abbott. *Flatland* was first published in 1884 (in London), and it's the only known example of a tale where the term "two-dimensional" is not a criticism but a factual description.

The premise of *Flatland* is that there is a flat, two-dimensional "world" inhabited by living, thinking, creatures in the form of geometrical shapes and lines. The story is narrated by a polygon known only as "A. Square," who is one of the residents of the flat world. The square (a lawyer, of course) helpfully tells us about what it's like to live in a world as wide and thin as a vast sheet of paper. He also learns what it must be like to be in a realm where there are fewer or more dimensions than the two with which he has lived his whole life. I think this story serves as a useful window into the notion of incompleteness, and an imaginative illustration of the limits of our ability to perceive the limitations of our own personal framework. So let's look at the world of dimensions with that in mind.

This strange and difficult proposition of incompleteness comes first to the square in a dream in which he visits Lineland, a one-dimensional world consisting only of a line. In the dream, the square meets the ruler of Lineland. Square tries to explain to this regal but linear person that he comes from a world with an additional dimension, but the monarch can't or won't accept such incomprehensible heresy and the dream ends. Square then meets a sphere, a visitor from Spaceland, a three-dimensional world (like our own). This sphere is something of a prophet, returning to Flatland at the turn of each millennium to appoint one apostle among the Flatlanders who will try to persuade the others of the reality of Spaceland and a third dimension. (And my mother thinks *I* don't visit often enough.)

In Flatland, Mr. Square only perceives Sphere as two-dimensional (a circle) because that is all a two-dimensional perspective can do with a three-dimensional sphere. Think about it—imagine taking a very, very thin slice of a hollow round ball and seeing what it looks like in two dimensions. But the circle/sphere then transports Square to Spaceland where he can literally look at things from a different angle. Once in Spaceland, Square can see for himself what that additional third spatial dimension means, and how it changes the objects and processes he had thought he'd comprehended before. The poverty, inaccuracy, and misleading simplicity of Square's Flatland point of view become clear, and Mr. Square's mind is opened to the startling new concept of multi-dimensional realms.

Square, thus enlightened, tries in turn to persuade Sphere that there could be further dimensions beyond the three of Spaceland, maybe even many more. Ironically, the sphere (the mind-bending prophet of three dimensions) can't wrap his curves around the idea of any dimensions higher than the three familiar to him. He feels insulted by the very notion. Sphere angrily returns Mr. Square to Flatland where he earnestly but unsuccessfully attempts to persuade his shapely but flat comrades of Spaceland's existence. The intrepid attorney, thus inspired and disillusioned, moved and humbled, realizes the widespread (maybe universal) phenomenon of ignorance and incompleteness that afflicts people confined to any Gödelian system of internal consistency and axioms. He grasps the incompleteness of Lineland when compared to Flatland; of Flatland relative to Spaceland; and possibly of Spaceland when juxtaposed with still higher-dimensional realms.

Like a prophet without honor in his own land, Mr. Square enjoys no more success in preaching the reality of three dimensions to his fellow Flatlanders than he'd had when reasoning with the sphere about the prospect of additional dimensions beyond his own three. In fact, the leaders of Flatland announce a new law imposing severe penalties for anyone caught advocating the false doctrine of three dimensions, and Square, like other visionary lawyers before him, is ultimately imprisoned for his heretical activities. Mr. Square knows the truth, though, and steadfastly refuses to renounce his troubling beliefs, even after years of confinement. It seems that once you recognize the validity of the incompleteness theorems, the most famous dictum of the great author Thomas Wolfe snaps into effect with full and unforgiving force: You can't go home again.

Lest we be too quick to judge the biased, excessively rigid rulers of Flatland, let's think about how reality appears from their perspective. For a person in a two-dimensional planar world, all shapes appear to be either dots or straight lines, regardless of their "true" contours when viewed from a third dimension above (or more precisely, orthogonal) to the plane. In a realm with no real thickness (just height and width), anything that is part of your plane of existence has no reality other than the edge nearest to you. If a straight line segment approaches you straight-on and head-first, like an arrow, all you'd experience is the small dot-like cross-section of that line segment, which would look to you like just a narrow point. But if that same entity comes at you more broadly from the long side, you would see it more accurately as the line segment it is. In Flatland, by the way, this is the shape of females—no multi-sided figure, no curves, just a little straight, flat piece. I must remember to add the three-dimensionality of women to my list of grateful-fors when Thanksgiving next rolls around.

Strangely, though, in Flatland you would also perceive *any* shape other than the pointy-ended cross-section of a line segment as nothing but a line segment. Whether a figure is "in reality" (i.e., from our three-dimensional viewpoint) a square, triangle, circle, pentagon, rhombus, or octagon, the only aspect you'd be able to perceive visually from your constrained perspective in the same paper-flat plane as your fellow shape is his or her edgewise, side-view, 2-D apparent equivalent—a line segment. You might be able to gather clues from "feeling" the figure around its edges, or by asking your companion what shape he or she is in (Hey, dude, what's your shape?), but those are mostly indirect means of information intake. What you would actually be able to see would be artificially limited to a very incomplete, ultra-shallow, and flat-out misleading aspect of the whole person.

With the benefit of an added third dimension, a person in Spaceland can easily, even effortlessly see the entire shape of everyone in Flatland. In fact, it's much more impressive than just that. A 3-D observer is able to see into the center (the heart?) of every Flatland individual by gazing into his or her core from a bird's-eye view, and even to touch the insides. We might think of this capability as romantic, medically useful, voyeuristic, a rude invasion of another's space, or gross. But a Flatlander would only consider it the stuff of wildly unrealistic fantasy or science fiction. How can one solid polygon possibly look into the heart and inner essence of any other solid polygon, when they're all on the same plane, with no way of getting past the solid planar line segments forming a person's bodily exterior? For them there is no way even to conceive of such a thing, whereas it's entirely self-evident, take-it-for-granted obvious for a Spacelander. It's amazing how much opens up to us when we are free to go outside the limits of one incomplete system for additional sources of perspective and insight. The results might seem fantastically incomprehensible to those inside the solid box of the onion-skin-thin perspective, but they are revealed as anything but magical when we gain an out-of-the-box vantage point.

Buildings in Flatland have to be accessed through the always-open doorway, which are really just gaps in the exterior wall, because that's the only way in or out of a two-dimensional structure for a two-dimensional planar person who can only slide along but can *never* leave the plane to enter space by emerging either upward or downward. A solid line segment is a solid wall, a complete barrier in Flatland, because there's no option of passing over or under it. Any entirely enclosed shape (a complete square, circle, or any polygon) would be impossible for a Flatlander to enter or exit from within that incomplete system. A 3-D being could peer down at any Flatland structure and see right into the middle of it—even fully enclosed shapes like a rectangle—and could come and go from above or below, as well as deliver or remove objects, despite the impassable and solid two-dimensional perimeter walls. A 2-D hermit inside what she thinks is a hermetically sealed cell would have instantaneous company from any overhead Spacelander gazing down or parachuting in from above. But if you put Mr. Square next to the hermit inside one of those same no-exit shapes, he too would be a prisoner, literally boxed in because he would find only a solid wall blocking his path in every direction.

Flatlanders would be astonished by these sudden appearances and vanishing acts, as well as examples of wall-piercing X-ray vision, and would think of these 3-D capabilities as magical or miraculous. And indeed they would be inexplicable from within the confines of Square's planar perspective. Three-dimensional visitors would seem to have the power to see inside as well as appear and disappear from the innermost chambers of even the most multi-layered nest of fully enclosed cells, box within box within box, without any need for doors and without any difficulty whatsoever from all those solid, unbroken, flat walls. Such feats would be utterly impossible from a Flatlander's point of view as he or she slides and glides like a tissue-paper-thin hockey puck along the plane of the smooth, bump-free sheet that is their prison. And if a creative visionary Flatlander were somehow to imagine the true explanation for these evidently paranormal anomalies, there would be no way to prove it using the evidence and tools available in Flatland. The truth would still be true, but not provable—that is, not provable when relying on everything accessible from within Flatland's incomplete world.

When your entire universe is unalterably flat, there is no such thing as crossing a line (which, from my point of view as an inveterate line-crosser, sounds like an excellent way to stay out of trouble for a change). Any line without an opening in it may as well be a trillion miles high as far as Mr. Square is concerned, because he is absolutely powerless to jump over it or tunnel under it—something that would be ridiculously easy for Sphere or anyone else who happens to come equipped with a third dimension handy. The mysteries of boundary-defying transportation only become possible, provably realistic, and understandable when one departs Flatland's incomplete and limited system and gains the additional information available from a different perspective outside the Flatland system. Just as Gödel showed, an entire system (like Flatland) can be internally consistent, and can be governed by well-defined axioms (laws, rules, edicts, and principles), yet also incomplete...with true features that can never be proved true using only the resources the system supplies.

Similarly, Flatlanders possess no concept of the light source in their world, because they know nothing of what it means to rise above their frame of reference. From their incomplete frame of reference, "above" has no practical, comprehensible meaning; and they have no experiential basis for grasping how it could be possible or what it would involve. For them, light just is, and it's a wonderful and inexplicable mystery. No one knows where it comes from, and it's present whether one is inside or outside any building. A three-dimensional person such as Sphere (from Spaceland) would see that the light is shining down on the Flatland sheet from the Sun overhead, and that it shines onto (and into) everyone and into every building from above. This description of light—its origin and its nature—is true, but no one in Flatland could ever prove it. Again, this phenomenon is unavailable and inaccessible within the Gödelian limits of the incomplete but internally consistent Flatland system. Only from outside the system, and with the benefit of information impossibly unavailable within the system, can someone see the light. You have to get beyond Flatland to prove the light source's location or the reason why light shines everywhere at once in Flatland. Why is light in Flatland uninhibited by even the most complete and thickly-solid walls of the most secure and fully enclosed doorless and windowless vault? From inside Flatland you can guess and you can wonder at this mystery, but for proof you must break out of the box.

Gödel's insights can help us to understand the unconquerable obstacles confronting any inhabitant of Lineland when trying to describe, to find words to express, or even to imagine the nature of Flatland. Analogous and equally insurmountable barriers confine people in Flatland who might be bold and inventive enough to want to contemplate something like Spaceland, and citizens (like ourselves) of Spaceland when we screw our courage to the sticking place and attempt to conceptualize still higher-dimensional realms. I hope Gödel's incompleteness theorems become more accessible to you when you think about an internally consistent and axiomatic (rule-based) system like Lineland or Flatland, and how the limitations inherent within these incomplete systems prevent us from proving some true principles.

Our helplessness (in Lineland, Flatland, or even Spaceland) is worse than that of a fish out of water, because at least a fish can briefly leap into the air world, glimpse it with its own eyes, and feel for itself a fleeting sampling of what it's like to be there. But the incompleteness of our nice, consistent system absolutely guarantees that we have no way to take even the shortest field trip to a higher-dimensional realm. Someone in Lineland can only travel back and forth endlessly along that one single line that forms her entire world, with no power to move past anyone else in that same straight and narrow path. A line is marked by points on each end, and a plane cannot fully exist in two dimensions, so it's impossible for residents of Lineland to move freely within a plane. Lineland's inhabitants clog up like railroad cars forever on just one track, with no movement possible other than rolling along its single link. In a one dimensional world like Lineland, everyone has no option but to stay on track; there's no way out—no over, under, or to the side. (You thought *you'd* been in bad traffic jams?)

Likewise, a plane is marked by lines at the edges, and three-dimensional space can't adequately be represented in a plane. Because of this, beings in Flatland can only travel within the 2-D limits of their plane, and never in the higher realm of space that we Spacelanders know is out there if you can protrude above or beyond that sheet. We shouldn't be surprised that Square is baffled by what Sphere tells him about a third dimension. We're in the same predicament, hemmed in by the three dimensions that bound our incomplete but internally consistent existence. When everyone in our world is and always has been permanently blocked from visiting, experiencing, sampling, testing, or experimenting with a place, that dearth of interaction spells death for our capability to prove things about or from that place's perspective. We tend to lack even the basic tools of language, physical manipulation, and thought to comprehend such a forever-foreign scene or to portray it in a way that makes sense within our familiar frame of reference. It's literally out of this world.

The inevitable incompleteness of our system is understandable when we consider it from this angle. We've been unalterably deprived of any means of gaining access to higher dimensions than our own. This has the effect of impoverishing our repertoire of available words, measuring devices, concepts, ideas, tools, and tests in any situation involving those higher realms. This prevents us from proving, with the eternally and internally limited resources available to us within our realm, certain things about our own system that are in fact true, but that require the perspective of an extra dimension for their proof. We might come at the problem indirectly by resorting to analogy and metaphor, but whatever insights we gain from such creative approximation would fall short of actual proof. We ineluctably lack the appropriate descriptive words, mental images, mechanisms, and even the ability to imagine, any dimension beyond that of our own. Our home may thus be forever incomplete, but at least, like Flatland, it's a nicely consistent and well-ordered little system. Cozy, but incomplete. Maybe a realtor could make that sound more alluring.

Irrespective of whether such outgrowths of Gödel's incompleteness theorems are valid on a rigorous level, it can be interesting to consider the possibility: We might glean from them some insights into the greatest challenges in international environmental law. As with Pascal's Wager, Gödel's incompleteness theorems could move us to reflect on the fundamental uncertainty that seems inherent in the problems of climate change, a contemporary mass extinction, and other current threats. If doubt, indefiniteness, and incompleteness are in some sense inevitable features of all scientific theories and the means by which we test them, we should not demand an impossible level of certitude as a prerequisite to taking action.

In fact, it is worthwhile to consider the impact Gödel's breakthrough might have had on the work of Blaise Pascal, had it been available to Pascal at the time he was contemplating his famous wager. I don't think the introduction of any particular aspect of the incompleteness theorems would have moved Pascal off of his central focal point. I believe that, at least on some intuitive level, Pascal knew that truth is a more powerful concept than provability. He understood, as evidenced by the way he structured his wager, that within any consistent system of information, some things are true that can't be demonstrated to be true through use of the tools and materials available in that system. Pascal was convinced that belief in God or in God's wishes for us is not the stuff of strictly rational analysis, deductive reasoning, and empirical observation, but that nonetheless a rational and logical person can rely on probabilities and reasonable risk/reward analysis to draw sound conclusions. This doesn't change with the infusion of Gödel's incompleteness; indeed, incompleteness and lack of universal provability are in some sense taken as givens in Pascal's Wager.

The contribution Gödel's work might have made to Pascal's analysis deals more with the amount of doubt confronting any decision-maker, rather than the existence of doubt. Pascal may have realized that there are more variables in his equation than he had thought. Arguably, Gödel could have persuaded Pascal that there are more unknowns relevant to his wager than we can list or imagine. But even if there is greater uncertainty clouding our ability to see through a glass, darkly, that doesn't negate Pascal's main point. Pascal would still argue that a rational person in a fog of doubt would make the most significant decisions based on a careful weighing of the likelihood and dimensions of all possible outcomes, including very good and very bad extreme potentialities.

CHAPTER 6

Heisenberg's Long-range Forecast: Fog Forever

In this chapter, we move to our most elusive and enigmatic source of help. Our quest to find a safe path to the future through the darkest of moonless and foggy landscapes is a perilous one. The ground under our feet is both hidden and riddled with riddles. But when life gives you fog, you film a mystery movie. And no purer fount of both mystery and fog-piercing vision exists than the towering figure who gave the world the uncertainty principle.

Werner Heisenberg was born in Germany on a wintry day on December 5, 1901. His father, Dr. August Heisenberg, was a secondary school teacher and university professor who taught the classics. His mother, Annie Heisenberg, was privately educated; and her fluency in Russian made her a frequent recruit to translate research papers for her husband's use.[1] The family was raised in the Catholic tradition, but both parents allowed their sons to decide their own beliefs. As Werner recalled, "[M]y parents were far away from the Christian religion as far as the dogmas were concerned, but they would always stick to the Christian ethics. They would accept the rules of how to behave and to live, and say that we can take them from the Christian religion, but we cannot accept literally all these old stories."

Dr. August Heisenberg was a rigid academic, consumed by

restless ambition that took a heavy toll on his family. Annie was perfectly suited to handle her husband's fluctuating moods, and always had a pleasant demeanor and a balanced attitude. She favored Werner over the Heisenberg's other son, Erwin, likely because Werner had suffered a nearly fatal lung infection when only five years old. Severe allergies and illness would continue to plague Werner into his adult life, but he held steadfastly to his mother's qualities and always maintained a balanced disposition.[2]

From an early age, August encouraged Werner and Erwin to compete in an intense sibling rivalry. The conflict ultimately ended their relationship altogether when the boys got into a particularly violent fight. This competitive spirit was instilled in Werner so deeply that he carried it with him throughout his life, and constantly felt the need to surpass others in every undertaking he attempted. When he wasn't playing classical piano or the violin, Werner was perfecting some athletic feat, such as skiing or running. In academia, Werner's love of mathematics sparked his desire to dabble in a new field of study that presented a mountain of challenges he yearned to take on—quantum physics.

As a doctoral student, Heisenberg attended a lecture by the great Niels Bohr on quantum theory in Copenhagen in June of 1922. While sitting in the back of the lecture hall with the other junior scientists (just as my students do to this day), Heisenberg had the audacity to voice his objections to one of Bohr's principles, much to the chagrin and embarrassment of the senior scientists. After the lecture, Bohr invited Heisenberg to walk with him to a local coffee shop, where the two analyzed quantum theory. Heisenberg always remembered Bohr's encouragement to look beyond the elaborate calculations and cleverly drawn-out models to comprehend the real physics that lay beneath. Bohr stated, "[W]hen it comes to atoms, language can be used only as in poetry. The poet, too, is not nearly so concerned with describing facts as with creating images and establishing mental connections." "My real scientific career only began that afternoon," Heisenberg recalled.[3]

When Heisenberg returned to Munich to complete his doctorate in the spring of 1923, he was required to take a laboratory course in physics to be sure he had established mastery over both theoretical and experimental physics. Heisenberg only reluctantly obliged, due to his strong dislike for experimental physics, and he enrolled in a laboratory course under Professor Wilhelm Wien. Wien was a distinguished researcher who openly criticized quantum theory, and this happened to be the specialty of Heisenberg's instructor Arnold Sommerfeld. Whether it was because of Heisenberg's contempt for laboratory physics or Wien's eagerness to assail Sommerfeld's latest "wunderkind" during his final exam is unknown, but Heisenberg was unable to recount the textbook formula required to answer the exam question correctly, and Wien became outraged. It was only thanks to an intense negotiation between Sommerfeld and Wien that Heisenberg was given a passing grade on the exam. One of the most ingenious minds of the 20th Century (or any century) thus received his doctorate with a grade just a shade above passing.[4] Students everywhere can take some comfort from this, but shouldn't become complacent unless they are so well prepared they can equal Heisenberg in their overall mastery of their subject. Sorry.

Heisenberg's most troubling times by far were the seven years he labored as a physicist under Adolf Hitler and the Third Reich. In 1937, Johannes Stark, an accomplished German physicist who viewed Heisenberg as competition for scientific prominence under the Nazi regime, publicly criticized and degraded Heisenberg for supporting Einstein's Theory of Relativity, calling him a "white Jew." To save his career, his family, and his life, Heisenberg wrote a letter directly to Heinrich Himmler, the Reichsführer-SS himself. Heisenberg gave Himmler the ultimatum of either approving of the public attack by Stark, in which case Heisenberg would resign, or disapproving of the attack and thereby restoring honor to Heisenberg's name. Fortunately for Heisenberg, his mother personally knew Himmler's mother through a social club, and offered to give the letter to Mrs. Himmler to deliver to her son. Himmler responded and accepted Heisenberg under the terms that he would be investigated by the dreaded Schutzstaffel, or S.S.[5] Shortly after the investigation, Himmler stated, "I believe Heisenberg is decent, and we could not afford to lose or to silence this man, who is relatively young and can educate a new generation."[6]

Heisenberg continued to travel and lecture around the world during the years leading up to World War II, and many of his colleagues never understood why he refrained from leaving Germany, just as Einstein and Gödel had. Heisenberg was a true patriot who wanted his family to remain in Germany, and, perhaps to his detriment, always remained optimistic that the political tide would change and the turmoil would soon dissipate. While lecturing in Ann Arbor in 1939, Heisenberg was asked by Enrico Fermi why he had not emigrated. To that question, Heisenberg responded, "I would feel like a traitor .... I don't think I have much choice in the matter. I firmly believe that one must be consistent .... People must learn to prevent catastrophes, not to run away from them. Perhaps we ought to insist that everyone brave what storms there are in his own country."[7]

One month from the time Heisenberg made that statement, Nazi Germany was at war, and the whole world would soon follow suit. Years later when Heisenberg looked back on the time he served as a physicist of the German Reich, he explained the position that he and the other German scientists took when given the task of creating an atom bomb to be used as the ultimate weapon by the Nazis: "The German physicists had consciously worked from the very beginning toward maintaining control over the project, and they used ... their influence to direct the work."

A lingering question is Heisenberg's role in the German atomic weapon project. Did he do everything he could to bring Germany the game-changing, war-winning weapon above all other weapons, only to fail bitterly? Or did he secretly sabotage and misdirect his team's work to ensure that Hitler would never wield the world's most terrible sword? Entire books have been devoted to both propositions, with— maybe fittingly—no definitive answer emerging. Although critics of Heisenberg argue that he supported the Nazi party, others have suggested that if anyone in the world was capable of creating an atom bomb at that time, it would have been Heisenberg. Yet he and his colleagues never produced such a bomb, and they were working while burdened with the knowledge that it would lead Hitler and the Nazis to almost certain world domination. All we truly know for sure is that, under the direct leadership of perhaps history's most electrifyingly innovative physicist, showered with every imaginable resource and material advantage, and under intense pressure from desperate, ruthless men who knew how to apply pressure, the German A-bomb project was a dud. Heisenberg always felt that he preserved both his science and his morality during World War II. He referred to his involvement in the war effort as "passive opposition"—a modest posture required for survival given the circumstances.[8]

Heisenberg remained a powerful figure in the physics world almost until the time of his death on February 1, 1976 (from cancer of the kidneys and gall bladder). A few years before his death, he was basking in the Bavarian countryside, listening to classical music played by his sons. He later remembered his thoughts from that moment of contemplation, "[F]aith in the central order keeps casting out faintheartedness and weariness. And as I listened, I grew firm in the conviction that, measured on the human scale, life, music and science would always go on, even though we ourselves are no more than transient visitors or, in Niels' [Bohr] words, both spectators and actors in the great drama of life."[9]

Like Gödel, Heisenberg was a contemporary and associate of

Albert Einstein. Heisenberg and Einstein fed off one another's ideas to create some of the most revolutionary advances the study of physics has ever known. Heisenberg made remarkable contributions to such daunting and preternaturally significant fields as relativity and quantum mechanics.[10] Quantum mechanics[11] itself has vast implications for our understanding of reality and is far beyond the scope of this book, dealing as it does with some of the most extraordinary and fundamentally powerful features of the physical universe, down to phenomena at the atomic and sub-atomic level. But Heisenberg's single most famous achievement, known even by many who have no idea what it really means, was his "uncertainty principle."

Published in 1927, Heisenberg's uncertainty principle, in its simplest form, holds that we can never simultaneously precisely determine both the position and the momentum or velocity of any particle.[12] In other words, as we move toward more definite measurement of a particle's position, at the same instant we can know that particle's momentum or velocity with ever less precision, and vice versa.[13] The uncertainty principle also limits the accuracy of simultaneous measurements of energy and of the time needed to measure that energy.

The uncertainty principle is often confused with the related concept of the "observer effect," which states that our very act of measuring a physical phenomenon, such as a particle's position or momentum, inevitably and unavoidably changes that which it measures. According to the observer effect principle, the observer or experimenter becomes part of the phenomenon being studied, and exerts an influence on it. No observer can be truly objective, neutral, or separate from what is being examined, therefore.

For the discipline known as quantum mechanics, the uncertainty principle is a key component. The uncertainty principle in turn depends in large part on the strange-but-true discovery that some, and perhaps all, matter has some characteristics that are consistent with how waves behave, and also other characteristics that are akin to the way particles behave. Quantum mechanics has shown this wave-particle duality is genuine and observable, most famously for photons (tiny "packets" of light) but for other, larger entities as well, such as atoms and molecules.

In fact, some versions of non-relativistic quantum mechanics contend that literally *all* matter exhibits the wave-particle duality, even on a macroscopic level, but that we can't observe the wave function of big objects because the wavelengths are so short. We won't wade too deeply into the muck of these murky marshes in this book, but suffice it to say that this startling principle has been verified experimentally (as with demonstrations of the photoelectric effect, for example) and it's no joke, nor is it some new-age mystical quasi-religious doctrine.

In a way, the wave- or particle-like features we see depend on the aspect of the thing at which we're looking. It's somewhat along the lines of the old fable from India about several blind men, each of whom touches only one part of an elephant and offers an opinion on what type of object he is feeling. If you limit your observation to only an ear, you might describe an elephant as like a large, flat fan, but if your arms are wrapped around one of the pachyderm's legs, you would be more likely to say it resembled a massive pillar (which would make a lousy fan). Yet an elephant is neither a fan nor a pillar…it's an elephant, with some characteristics in common with both a fan and a pillar. Likewise, a photon is neither particle nor wave…it's a photon. But it has a wave-function and it also has some features that it shares with particles.

The effects of the uncertainty principle on precision may be real, but infinitesimally small, for objects large enough to be part of our ordinary human experience. However, for extremely tiny, submicroscopic objects such as atoms and subatomic particles, the uncertainty principle produces significant and measurable results. One of the most important ramifications, within quantum mechanics, is that the uncertainties at the level of atoms and subatomic particles are sufficient to reduce our ability to ascertain their velocity, position, or momentum to an exercise in statistical probability. It is impossible to make entirely accurate predictions about cause-and-effect phenomena at the particle level because of the unconquerable limitations the uncertainty principle places on the simultaneous measurement of time, energy, velocity, and position.

The uncertainty principle, under modern interpretations of Heisenberg's innovations, is a property of so-called quantum states which correspond to the statistical features of the measurement process. In a foolhardy attempt to fathom the unfathomable, allow me to outline a commonly used thought experiment that might help to illustrate what this means. Imagine that a physicist has a way to obtain and examine an electron in some particular quantum state. Let's see what the uncertainty principle says about repeated experiments involving this electron. Assume that the physicist very carefully replicates every detail of this procedure 200 times. Suppose that for the first 100 times he or she measures the electron's position, and for the second 100 times he or she measures its momentum or velocity. Irrespective of the quality and precision of the super-accurate measuring apparatus, the measurements actually taken will be different in every single one of the first 100 and second 100 experiments. The readings will tend to cluster around some mean (average) figure, with some variation around that mean, as measured by the standard deviation; but each one will nevertheless be different from all the others to at least some small extent. In this way, the electron's actual position and momentum/velocity can never be precisely and simultaneously ascertained; rather, they will reflect a probability distribution, with various combinations of both attributes being more probable or less probable.

It is crucial to grasp that this inability of ours to measure precisely both the location and momentum or velocity of a particle at the same instant is not, repeat *not*, merely a function of inadequate equipment. Heisenberg's insight is that even with theoretically perfect, tricked-out, top of the line, fully loaded measuring devices—beyond the quality of any we could ever imagine creating at any price and at any point hundreds of years of technological advancement into the future—there would still remain the fundamental and ineradicable uncertainty inherent in any attempt to simultaneously pinpoint the two dueling aspects of our annoying particle. It's not that our equipment sucks. It's something built into this strange system we call reality that sucks the certainty out, no matter how state-of-the-art our equipment may be.

This is difficult for many of us to understand, because we've been soaked and saturated all our lives with breathlessly hyped advertisements for products from aardvark aspirin to zebra zippers. We've been carefully taught that there is no problem that an upgrade to the latest bells-and-whistles version 7.03 techno-tight marvel won't fix. Just add money.

But take it from Heisenberg and try another thought experiment, this time with you as the guest star. Imagine yourself transformed to a sub-atomic level (through the wonders of liposuction gone too far, maybe?) with the fascinating job of traffic cop for electrons. You should've listened to your mother when she told you to go to law school. The only advantage of the situation from your diminutive perspective is that a single normal-size doughnut is sufficient to feed you and your entire police force for a week. But there you are, waiting in the fog to issue speeding citations to any electron that exceeds the posted speed limit in a particular neighborhood.

To help you do your job as the smallest law enforcement officer in history, you have been issued the finest, costliest, next-generation combination radar gun and laser beam equipment conceivable. You dutifully do what you have been meticulously trained to do; you aim your mechanical marvel at every suspicious-looking electron that whizzes past your hiding place and take the best possible readings.

And you are frustrated! Despite your flawless operation of the perfect measuring device, you find that your results are never 100 percent accurate for simultaneous readings of both the location and the velocity of any nano-NASCAR electron at any given wink of time. Close, but no butterfly net. Again, it isn't the equipment's fault, and it isn't that you're using it incorrectly. There is literally nothing anyone can do about it. As political wonks might be wont to say (if they knew Heisenberg's uncertainty principle from the Heisman Trophy), "It's the uncertainty, stupid!"

The uncertainty principle, like Gödel's incompleteness theorems, can spawn some very far-reaching ramifications, whether or not they are properly within the original insight. Because, according to Heisenberg's uncertainty principle, certain facts are not objectifiably knowable in quantum theory, there could be profound implications for the epistemology of science in general, i.e., how do we know what we know, and how do we know what we don't know.

Before Einstein and Heisenberg, physics was viewed as an exact and empirically verifiable, experimentally observable science. Back in the day, that breed of physics was in evidence all around everyone, and it followed mathematically predictable rules as described by Isaac Newton. Now, with the advent of the uncertainty principle, the outcome of a physics experiment is understood as depending on the view of the observer. If such uncertainty is wedded to the "exact science" of physics, what are the implications for less formalized branches of human knowledge? Logically, it appears to follow that there will be different valid interpretations of the same phenomenon, and that these interpretations will often be in conflict with each other. Taken to a logical extreme, this could mean that we can never know the ultimate truth. The philosophical principle of indeterminacy is derived from this source, whether or not that's strictly called for when we're playing by Heisenberg's rules within his specific area of concern.

As you might imagine, the intricacies of the uncertainty principle are beyond the scope of this book (and, let's face it, beyond the powers of this author). Suffice it to say that Heisenberg's innovation was intensely disturbing to many life-long scientists, even the great Einstein. The idea that an electron does not follow a precise orbital path, but rather exists within a range of positions and velocities described by a probability distribution, provoked Einstein to object, "The Old One (i.e., God) does not play dice!" If Heisenberg transformed Newtonian physics into an elaborate game of chance, what room remained for certainty anywhere in the universe?

Heisenberg's work on uncertainty, with a strong assist from Gödel's incompleteness theorems, has distant and not-so-distant relatives in modern disciplines very different from the subatomic level on which quantum mechanics concentrates. One example with particular relevance to any discussion of climate and the prediction of weather patterns is the work of Edward Norton Lorenz and others on chaos theory and the butterfly effect. Let's take a brief side trip from Heisenberg's immediate neighborhood and see the chaos that ensues when we let a universe full of uncertain and unruly butterflies loose onto an unsuspecting world.

Lorenz, an M.I.T. meteorologist, found in the early 1960s that mathematical models for predicting changes in atmospheric convection, such as in the forecasting of weather, would yield significantly different results when run multiple times or using extremely minor variations in starting conditions. Similar to Heisenberg's insights on probability distributions instead of precise, discrete measurements of an electron's position and momentum, Lorenz discovered that the way in which air moves in the atmosphere could not be entirely and precisely predicted using his twelve-equation computer model. Instead, the startlingly counterintuitive finding was that even the tiniest, most minute variations in the initial values and conditions could result in major divergence in final output. Our instinct would generally make us assume that, for example, a one percent increase in the initial value of a given meteorological variable that we put into Lorenz's computer simulation would yield something along the order of a one percent shift in the output. That would make sense, and would be in accordance with our everyday real-world experience. And it would be wrong. Very wrong.

In fact, Lorenz himself found that immense differences in outcome could flow from using a starting value for one variable of .506 instead of .506127. Even though a change of about one part in 10,000 is extremely difficult if not impossible to measure accurately, such a minuscule adjustment of initial conditions can produce tremendous variation in eventual output. Who would have thought this possible? Hey, even someone as cheap as I am will tell the taxi driver to keep the change when the fare is $99.99 and I hand him a one-hundred dollar bill. That one penny, amounting to 1/10,000 of the whole tab, is such a tiny fraction of the whole that it seems not to be worth quibbling about even for the most stingy, penny-pinching tightwad. Come to think of it, the fact that I give tips of this size may be why I have so much trouble getting cabs to come to my house. I might view this as chaos theory in action, or merely payback for my lack of payout.

The stunning revelation from Lorenz's observations was that, even for relatively simple systems with only a few variables, we can't really predict the results that flow from the smallest hiccup in initial values. Lorenz concluded that it was impossible to predict the weather accurately, but then I could have told him that much just from my sorry experiences growing up in Chicago, wondering whether any particular Cubs game would be rained out or snowed under by the time I arrived at Wrigley Field. Eventually, though, Lorenz and others took the analysis much further to a broad array of subject matters. Chaos theory, or the study of sensitive dependence on initial conditions, became a way of attempting to find the order hidden within seemingly random or chaotic phenomena.

The butterfly effect[14] is the common nickname by which we know

Lorenz's discovery that "complex systems" can be exquisitely sensitive to the smallest variations in initial conditions, with widely divergent results based on multiple trial runs. Like Heisenberg's electron cloud, the butterfly's wings represent a broad range of different outcomes that a complex system such as weather and climatic patterns can produce based on slight adjustments in starting values, or even multiple repetitions using the very same values and conditions. It is important to this book's thesis because it reveals the limitations on our ability to control events. We can't confidently predict how any particular change will manifest itself, whether that change comes from natural forces or from our own human intervention.

The butterfly effect cautions us that the same net will not always catch the same butterfly under the same circumstances. If we repeat our efforts many times, we won't always obtain the same results, regardless of how meticulously and precisely we control for variations. And once we factor in the differing starting values and variable conditions that inevitably complicate our enterprises in reality, we can expect even more dramatic fluctuations in outcome from one occasion to another. Especially when we venture out of the spotless, self-contained laboratory and try our hand at experiments entailing entire continents full of living things and real weather. Then Heisenberg's uncertainty principle swoops down to beat us with the wings of a million million butterflies. Theoretical elegance can easily get lost in a vast cloud of dust—or the tiny, colorful scales flying loose from all those myriad fluttering butterfly wings.

If we can't even precisely predict the local weather, what hope do we have of forecasting the future of global climate change, with or without our interference in the variables of nature? Even if we summon all our resources and make a full commitment to solve an environmental problem, there is no guarantee that our actions will pay the dividends we desire. This uncertainty, even chaotic uncertainty, is a prime reason why there is genuine scientific debate as to cause of, magnitude and immediacy of, and optimal solution to potentially devastating environmental hazards like a contemporary mass extinction. In our quest to save real butterflies, the butterfly effect might toss our well-intentioned efforts into the dustbin of unintended negative consequences.

With apologies to Heisenberg and Lorenz, I want to interject some personal thoughts that their icon-shattering work has inspired in me. These meager ideas have no proper place alongside their epochal, veil-ripping breakthroughs, and they pertain to a very different field of endeavor from the subjects they transformed. Still, this is my book, and if I want to veer off onto a marginally-related tangent, I think that is my prerogative. It's sort of like those bad movies that begin with the tantalizing message, "Inspired by a true story" and then proceed to tell a tale bearing virtually no resemblance to anything that ever actually happened. Just bear with me for a moment, okay?

In my somewhat perverse, personal opinion, Heisenberg's work on uncertainty and Lorenz's chaos theory help illuminate something peculiar about the way people make decisions. A little earlier in this book, in Chapter 4, I discussed decision theory and a few of the models that try to show us some rational, logical methods of making choices. There are choices to make even among these choice-making methods, but all of them share a devotion to reliance on evidence, reality, reasonableness, and logic-based thought in arriving at our decisions. They reflect a commitment to rationality as the proper guiding force behind our efforts to choose wisely in the most significant judgment calls we face in our lives.

However, there is an entire branch of decision analysis that I left unexplored in Chapter 4, and that deals with the study of how people *actually* make decisions rather than how they *should*. Maybe this distinction occurred to you while you were reading about the various analytical frameworks available for rational decision-making. Maybe the thought came to you, "Funny, but I don't think that's how most people really make their biggest decisions." You might have pondered, "I don't think I have ever known anyone who actually used a formal logical framework to cut to the core of any major dilemma, from proposal of marriage to choice of profession to the yes-or-no of procreation." These are valid points, and ones that have not eluded the pioneers in the field of decision theory. In real life, there is often a huge gap between reality and theory, and between what should be and what is.

What accounts for the widespread human tendency to make irrational and even objectively self-destructive decisions? Why do people embrace "Thorn Birds" decisions[15] in which they more or less know that terrible long-term consequences will likely flow from a particular short-term choice, yet make that choice anyway? Why do people "go with their gut," play a hunch, or opt for emotional/sentimental alternatives, even when faced with what appears to be incontrovertible evidence that these decisions are far from rationally optimal? And is there any cure for this tendency to make illogical and wildly counterproductive choices?

I was tempted to list some notable examples of irrational, objectively ruinous decisions from the recent past that anyone with a modicum of common sense could have predicted would inevitably lead to disaster. But then I realized that such a list could easily and swiftly expand into a 37-volume *History of Humanity* magnum opus, and at this stage of my book, I'm inclined to keep my opus on the non-magnum end of the literary spectrum. If you want a collection of for-instances in which human beings have made horrible choices in defiance of the obvious facts, probabilities, and logically-inexorable outcomes, you can obtain an up-to-date batch by glancing through today's news stories. This is the ultimate inexhaustible and endlessly renewable resource.

I've written this book to set out a way for all of us to make logically correct, evidence-guided, probability-based, rational decisions regarding the future of our planet. I've devoted a lot of work to making this argument as reasonable, accurate, and rooted in sound logic as possible. Yet I am enough of a realist to know that at all levels of human activity, including the national and international sphere, rational decision-making is most definitely not the usual practice. From the schoolyard playground to the world's capitals, decisions small and momentous are often made from a very different assemblage of raw materials. When we're deciding on our choices, we frequently don't reach for objectively-verified evidence, rigorous probability analysis, reasoned cost-benefit calculus, or empirically verified experimental results. Instead, we rush to an all-you-can-eat buffet overflowing with prejudice, powerful emotion, wild guesswork, untested assumption, shortsightedness, laziness, convenience, long-entrenched habit, and desire to follow the herd.

I understand that people are not robots, automatons, or Mr. Spock clones ruled by reason and virtually free of emotion. I know that, and I am not really arguing that this is inappropriate at all times and for all decisions in life. Some choices, especially when it's a close call and we need something to break the tie in favor of one course of action over the others, are well-suited to a healthy injection of love, fear, loyalty, hope, faith, and other powerful but perhaps non-logical forces. If we're deciding whether to marry someone special to us, or to have children, a decision matrix loaded with probability and evidence lacks a certain (or uncertain) something extra that we intuitively feel is necessary and important. There is more to life than numbers, science, hard facts, and logical rigor.

But many times we take this exception and run amok with it. It's very easy to go from acknowledging some proper and limited role for emotion and intangibles in decision-making to full-blown, out-of-all-proportion, wanton guesswork. For vast multitudes of people, decision analysis reflects very little involvement by the part of the body we call the brain. Instead of the brain, our choices are typically fueled by those anatomical features known as the heart, the gut, or the genitals.

Potent impulses burst forth from our visceral core. Hormones matter. Emotional powerhouses like love, sexual attraction, hope, yearning for immortality, fear, and fealty are major factors in making life worth living. No one could ever remove them from human decision-making, nor should we want to try. The problem, of course, leaps out from the shadows to bite us when we make our biggest and most widely impactful decisions in blatant defiance and disregard of objectively clear warning signals. With disaster looming, disaster is frequently what actually ensues in the aftermath of raw and unrefined, illogical, hormone-hijacked choice selection. But how big a role for intuition is enough? How much is too much?

Given the inevitable incorporation of emotion and intangible factors in our decisional activities, a sense of balance is crucial. The myriad devastatingly-bad decisions we've made as nations, neighbors, and nerds are the predictable consequence of allowing logic and rationality to be squeezed out of the equation entirely. Heisenberg teaches us that some amount of uncertainty is unavoidable at the subatomic level of quantum mechanics. Lorenz adds that our inability to define with complete precision the true value of all the relevant variables means the outcomes will reflect the butterfly effect in their unpredictability. But there is a vast difference between acknowledging the impossibility of completely eradicating uncertainty and imprecision from our decision-making (despite our best efforts), and savagely embracing the overwhelming predominance of gut, heart, and gonads over the brain. It's the distinction between reluctantly accepting a certain unavoidable level of uncertainty and risk inherent in the situation, notwithstanding all we can do to rid ourselves of these negatives, and failing even to recognize the problems posed by that uncertainty and risk...doing nothing to counteract them.

If illogical thinking is a widespread phenomenon all the way from the U.N. to the you-and-me, it's worth trying to figure out why. The prevalence of fallacious approaches to problem-solving is a formidable impediment to the central purpose of this book—to develop a plan to help people at all levels arrive at their own independent, soundly reasoned conclusions on what the world needs from its inhabitants. That purpose will not matter if we make our decisions on the basis of ABL..."anything but logic." Unless, that is, this is not a permanent condition, but rather one that could be changed given the right approach and the proper set of circumstances. So let's think of a few reasons why many of us might avoid rational, logically rigorous, analytical decision-making even in extremely weighty personal issues that directly and powerfully affect our own lives.

In part, the answer may be that we haven't been taught *how* to think. That isn't as much of a tautology as it might seem to be. In the modern educational curriculum, we rarely teach our students formal logic or logical reasoning. Even at the university level, we usually don't prepare people to recognize any of the numerous forms of logical fallacy.[16] We fail to teach young people about this menu of mistakes, these time-tested pitfalls in argument and persuasion. As a result, unless they study it independently or somehow figure it out on their own, they never grasp the trouble spots that can trap the unwary along the path to making good choices. This is a real problem, because these logical fallacies have been studied for thousands of years for a very good reason—they have been the main obstacles to wise decision-making ever since people first began to knuckle-walk the planet. For millennia we've repeated these same mistakes, countless times, in every age and every culture. Those who are aware of the fallacies have a fighting chance of spotting them and resisting their false charms. But if we forget these age-old lessons now, we'll be unprepared to handle the greatest challenges of all time. If we've ever needed to be equipped to identify and neutralize errors in logic, that time is now.

If you're one of those folks who never read footnotes, make an exception just this once and look at some of the common errors in logic that have plagued human beings since the dawn of time. Then look at how people stumble over such well-known logical tripping-points every day—and this includes supposedly well-educated, mature adults in leadership positions. Hey, let's choose up sides and play a little game. Are you ready?

Take any subject you know well, and then read a few blogs, op-ed pieces, or magazine articles that discuss your topic. How many logical fallacies can you find in the first five minutes? Do you see any *ad hominem* attacks, trying to undermine an idea because of supposed personal flaws in the idea's spokesperson? Can you spot a *reductio ad absurdum* in which someone attempts to destroy an argument by misstating it or its conclusion in a ridiculous, exaggerated, or absurd way? Are you able to identify an example of *post hoc ergo propter hoc* where the author argues that one thing must have caused another simply because it happened first? I'll bet you can collect a few instances of *argumentum ad populum* that attempt to persuade by claiming large numbers of people believe a certain thing to be true. You can probably identify some examples of an *appeal to authority*, too, where someone argues that an idea is correct because a famous or respected person has accepted it. Look at my footnote again, and keep searching for examples of these and other errors in reasoning. The list of fallacies goes on and on!

If you played this little game, even for a few minutes, you now know more about logical fallacies than a lot of well-dressed, well-paid individuals who sit in big offices making big decisions that affect your life. That explains quite a bit about why conditions so consistently suck all around the world. Heisenberg would understand, because he saw the fogginess unavoidably shrouding the tiniest entities in existence, and he knew from years of struggling to persuade some of the most brilliant minds in science that we add plenty of our own entirely avoidable uncertainty to everything we touch. Ignorance is bliss, and it blissfully multiplies life's uncertainty wherever it is unstopped by sound logic.

If you think about the dirty-laundry list of logical fallacies, you can probably discern some explanation for why they have been so persistent and broadly shared throughout the expansive centuries of human existence. The bottom line: Most if not all of the common fallacies are nice little mirrors on human nature's enduring aspects. They reflect our desire to be accepted and respected, to be part of the group and not isolated, to be on the side of dominance and strength, to take shortcuts, to search for easy and quick answers, and to feel superior. All of this is in addition to the fact that many of the fallacies involve a level of sophistication and precision of thought that is foreign to most of us. False logic can deceive us because we have no idea how it masquerades as the truth in various long-proven standard disguises. We simply don't like to think that hard or that carefully… it's too much work. We've never been asked to do it very much, if at all, and we're not really that into taking it up as a new practice now. And even if we are *willing* to take a hard look at our decisions, we lack the education and training to be fully *able* to discern the best options. No one has ever told us about the logical fallacies, and what we don't know *can* hurt us. It all adds up to a recipe for a nice, lengthy hit-list of stinking thinking that has stood the test of time through every age of humankind from Stone to Silicon.

In addition to our ignorance of the logical fallacies, as well as the tools and ingredients of proper logic, laziness is another significant reason why we tend to rely on illogical means to formulate our opinions. It's just easier and more expedient to depend on hunches, hasty guesses, reflexive reactions, untested assumptions, snap judgments, sloppy short-cut reasoning, and vague intuition. Winging it doesn't demand any structure, discipline, formal examination of evidence, determination of probabilities, or meticulous listing of relevant variables. Gut reactions require no more thought than anything else that involves expelling waste products from our alimentary canal. What could be more elementary? Unfortunately, the quality of the decisions that come from relying on our gut usually resembles the solid waste that issues forth at the end of our food's long, winding journey through our intestines. This type of gut check is about as rigorous as the ancient pagan practice of seeking guidance from examining the entrails of a sacrificed owl. As a tool for making wise decisions, one species' gut is as fetid as another's, whether owl or human. Yet an appeal to the intestinal tract remains a very popular decisional mechanism. It's enough to give you ulcers.

To be fair to the multitudes that resort to their gut, heart, or genitalia rather than risk straining their brain on life's daily dilemmas, those other organs can actually serve as sort-of surrogates for rational analysis, within many routine and familiar circumstances. At some general, subconscious level, we tend to be aware of most of the major factors, odds, and possible outcomes involved in the big decisions we face. We often have at least an approximate, ballpark degree of appreciation of what we're dealing with in the predictable, widely experienced problems that come with the territory of being alive. Without explicitly and directly subjecting our options to rigorous rational analysis, our subconscious mind possesses much of the relevant information, in rough outline and to a rough-estimate extent of precision. We can call it instinct, intuition, or educated guesswork, but this background-level mode of problem-solving is frequently sufficient for sifting through our choices.

Unless there are factors at play of which we're unaware, contingencies we've failed to recognize, hidden flaws in what we believe to be true, or unknown variables we've never heard about, our visceral reaction can be close enough for many of life's decisions. Our gut and our heart can supply us with approximate results that serve as pretty decent stand-ins for the hard work of structured logical analysis much of the time. We make a lot of instinctive judgment calls that turn out okay, so we get used to relying on those estimates. This is only a problem when we try to get by using this same imprecise holistic approach in those unfamiliar, unique, and/or troublesome situations where visceral big-picture approximation is inadequate to cover all the relevant intricacies. When close isn't close enough, that's when the brain needs to come to the rescue of our other decision-making organs—if we are prepared to recognize those situations and respond accordingly. That's when an unbalanced reliance on our irrational instincts—letting our head be ruled by our heart, gut, or sex organs—can leave us with major-league headaches, heartaches, indigestion, and decisional impotence. Ouch to the fourth power!

This victory of the viscera isn't merely a matter of our failure to think clearly about what to do at key crossroads. At times, there is an internal battle that rages within us over how we handle the problem at hand. We may actually use our brain logically and rationally, to formulate the objectively best course of action...and then choose to act otherwise, in knowing disregard of what's best! Even though we *know*—on a dispassionate, intellectual level—that we should select Option A, we *feel*—on an overwhelming, emotional level—that we must passionately embrace Option B. In these situations, our reason-rooted ideas do battle with our instinctive feelings. An internal conflict between head and heart, or between brain and groin, can be an intensely agonizing struggle. We're torn, from the inside out, between knowledge and emotion, and regardless of which "wins" we may feel as if we lost. It's a civil war in miniature, waged and raged within ourselves, and there is always a heavy toll of collateral damage.

When our lower organs clash with and conquer our brain, the decisions that result can be the most painful, harmful, and even tragic turning points of our lives. We fall in love with someone with a long history of instability, criminality, and violence...and follow this person down the darkest alley. We become sexually attracted to a person who should be off-limits because of marital status or professional relationship...and risk everything to yield to our lust. We crave the material pleasures we could buy if we only had a lot of money... and succumb to the impulse to steal from our employer. We recoil from the pain and loneliness of life, longing for a miracle cure that would stop the hurting...and surrender to drug or alcohol abuse. We hunger to fill the endlessly unsatisfied emptiness we feel inside ourselves...and give in to the urge to overeat, again and again. In each of these little dramas, and countless others, we understand that reason points us out the door, but we relinquish reason and rush in anyway. If something like this has ever happened to you, you have plenty of company. Thorn Birds of a feather flock together.

Many of us are very comfortable with making most if not all of our choices on the basis of emotion, easy cheat-codes, and guesswork. We've established deep, lifelong habits of thought. If this is the way we've always made our choices, even for life's biggest questions, it will take some vigorous nudging to move us out of that well-worn comfort zone. It will feel unnatural, mechanical, impersonal, and robotic to begin formulating decisions through use of a formal, logical structure after so many years of hormone-fueled judgment calls. Some of us may even feel—and I do mean feel rather than think—that it is inappropriate or morally wrong to be so methodical, at least where fundamental life issues are at stake. In a context rife with life and death, it can feel coldly calculating to handle the situation with sterile tongs instead of kid gloves. And in a sense the rational, methodical take on decision-making truly is literally calculating. It entails careful, meticulous, fact-founded calculation of the odds of each contingency happening, together with a disciplined attempt to quantify the magnitude of each possible outcome, good or bad. Such a quantitative, probability-related framework is indeed calculating.

The calculating mode also depends on another necessary component of rational decision-making often in short supply—the capacity to see and comprehend long-term and intermediate-term causes and effects. Commonly, people tend to be short-sighted and overly focused on short-term or immediate actions and their consequences. We're impatient and highly limited in our perspective. We see today and perhaps tomorrow on our radar screens. Anything beyond that scarcely warrants a moment's thought. We probably wouldn't admit to it, but unless a problem virtually shrieks with urgency and imminence, we sleep right through it. Unfortunately, a lot of decision analysis depends on identifying and assigning appropriate values for variables and contingencies that may happen—if at all—way off in the future. A point of view locked into the pursuit of immediate gratification and the avoidance of imminent destruction is poorly situated to incorporate in any realistic sense all factors that lurk in the realm beyond the universe's emergency room doors.

For people who tend to hate math in all its varieties, and who are untrained in the proper use of logic in evaluating the merits of competing options, it is natural to feel little attraction for the rational analytical model. Math and logic feel like the classes we hated most in school—the ones we tried to escape at all costs. And it's true, logically rigorous and rational problem-solving isn't easy. It insists on a fair amount of hard work and concentrated thought from us. To do it correctly even requires some significant preparation as we gather facts, weigh evidence, and identify the relevant factors at various points in time. That sounds a lot like homework! Is it any wonder that most of us vote with our feet propped up, and just go with the way that asks nothing of us beyond how we feel?

If we often take the plunge (sometimes literally, as well as figuratively) in a given situation and make our judgment call based on a predominant mix of emotion, feeling, intuition, and assumption rather than organized, logical rationality, what are the results? Sometimes the outcome is acceptable, even wonderful. Many decisions don't matter very much one way or the other. Many issues don't require in-depth sophisticated analysis. But what about the ones that do? In our personal lives, why do we make devastating and self-destructive choices, such as to abuse drugs, to gamble excessively, to commit felonies, to stay with a life partner who hurts us, or to drive while drunk? Why do we betray those who love us, fail to save our money for long-term goals, incur huge unnecessary debt, sink our life's savings into a highly speculative stock investment or business venture, neglect our health, or abuse our children? Why do we join a violent religious cult, engage in unprotected sex with multiple partners, fail to buy insurance on our house, or choose as business partners our friends who are involved in alcohol abuse or criminal activity? And why do nations get involved in ruinous aggressive wars, costly trade battles, divisive alliances with terrorist groups, religious persecution, political oppression, discrimination against various disfavored groups, and environmentally destructive activities?

In many of these objectively bad decisions, on both the individual and the collective plane, the chief problem is decision-making with not enough of a component of rationality and sound logic, not an overabundance of them. Conversely, there is rarely a shortage of unrestrained emotion, unthinking impulse, sloppy acceptance of fallacious analysis, and reflexive instinct to be found among the wreckage of these devastating choices. The bombed-out road to Hell is littered with used-up and worn-out logical fallacies.

It is largely this near-universal human penchant for making decisions based on everything under the sun except proper logic and reason that provides the foundation for the most internationally famous of all legal principles: Murphy's Law. Our wretched decisional technique is the main causal factor that underpins the oft-observed tendency for everything to go wrong that can possibly go wrong. In fact, in my opinion it may even be to blame for the existence of entropy, the accursed phenomenon whereby the amount of disorder increases within any system as time passes unless there's an input of energy and effort to counteract entropy's messy influence. Oh, all right, I admit that our flawed decision-making can't fairly be blamed for all the entropy in the universe. But to me it seems that, at least within the thoroughly banged-up sphere of human activity, our illogical and self-harmful decisions often team up with entropy to give us our perpetual state of crisis. This syndrome is so widely known that the military long ago made up an informal and irreverent acronym to describe it. Politely phrased, the term is SNAFU, which stands for "Situation Normal: All Fouled Up." Or words to that effect.

There. Doesn't it make you feel better to know that there's a quasi-scientific explanation for why we're continually busted by Murphy's Law? Me neither.

These last three chapters have established some principles that can lift us to higher ground above the terrain-smothering fog as we try to gauge the state—and possible fate—of the world. If we take some key insights and ideas from Pascal, Gödel, and Heisenberg, and think about how they can assist us in understanding these colossal issues, we can go far beyond the typical human approach to decision-making. We can minimize uncertainty and chaos, and gain some hope of—if not taming and domesticating them—at least confining all the butterflies and their fog-spreading effect within manageable limits. The remainder of this book is devoted to a detailed examination of what that quest entails, and how to pursue it without crashing into the sharp rocks hidden beneath the haze.

CHAPTER 7

Thinking Inside the Box with the World Wager

It is safe to say that most heads of state and most legislators remain quite unaware of the pivotal ideas from Pascal, Gödel, and Heisenberg. The same is very likely true for a solid supermajority of the world's citizens. The unavoidability of uncertainty in most if not all aspects of the world in which we live is an entirely foreign concept from top to bottom in every country on this planet, from pole to pole. As a result, when dealing with weighty matters of environmental and biodiversity law, the nearly universal default position is "When in doubt, sit it out." It doesn't take much to persuade decision-makers to do nothing, and there are usually some nice big question marks floating around the major environmental issues that serve this inertia-friendly purpose very well.

On the international law level, the problem has been serious enough and persistent enough that there have been formal attempts to redress it. The most obvious would-be curative is the idea known as the Precautionary Principle. The Precautionary Principle provides that, in order to protect the environment, the precautionary approach shall be widely applied by States according to their capabilities. Where there are threats of serious or irreversible damage, lack of full scientific certainty shall not be used as a reason for postponing cost-effective measures to prevent environmental degradation.[1]

In other words, if an issue is of great significance, and if an incorrect decision regarding that issue has significant potential to cause environmental harm that is sufficiently large and/or beyond remediation, the default option should not be to wait and see. Under this set of circumstances, the Precautionary Principle instructs nations to implement the more proactive "better safe than sorry" approach rather than doing nothing while awaiting irrefutable scientific confirmation of the crisis. As I have shown in the previous chapters of this book, full scientific certitude may be an oxymoron, and only a moron (whether "oxy" or otherwise) would remain forever inactive in the face of a major threat while looking for the certainty that cannot ever arrive. The Precautionary Principle is considered by some to have attained the status of customary international law,[2] and has been incorporated into several environmental treaties.[3] But, as is the case in many areas of international law, and indeed everywhere lawyers are involved, the meaning and effect of the Precautionary Principle remain open to differing interpretations.[4]

Lawyers and law professors tend to be more comfortable with uncertainty than people in most other professions, and this may partially account for the existence of the Precautionary Principle in some areas of international environmental law. In law school, students are often exposed to some extent with the various schools of jurisprudential thought, that is, the legal philosophical frameworks that attempt to explain and predict the way in which the law operates to address real problems. Two of the more widely discussed and most modern legal philosophies, Legal Realism and Critical Legal Studies, deal directly with the role of ambiguity in the law.[5] I will briefly describe how.

Legal Realism[6] emerged fairly early in the 20th Century as a response to the older, more rigid philosophies such as Legal Formalism. Legal Realism arose in reaction to the formalists' treatment of law as a species of scientific analysis with efficacious rules, logical processes, and rigorous pursuit of the demonstrably correct result in each case. The realists argued that law is not a branch of science, nor is it a precisely defined discipline along the lines of familiar forms of mathematics. On the contrary, Legal Realism posited that law necessarily includes elements of subjectivity, politics, prejudice, and an array of other very human attributes. The "realism" component of Legal Realism holds that it is inaccurate and simplistic to pretend that the law is above such influences, and that we can't properly understand how the law truly works without making room for the forces exerted by these wildcards added to the legal deck by human beings.

Critical Legal Studies[7] developed as a school of thought much later in the 20th Century, and is rather akin to Legal Realism on steroids. Critical Legal Studies reflects such influences as radical politics, civil unrest, postmodern thought, and the various branches of the civil rights movement. Adherents argue that political power, hierarchical repression, bias, wealth, and personal preference combine to produce legal results that represent a system dominated by uncertainty and human conflict. Critical Legal Studies suggests that rules do not decide cases…people do, and not always for good, fair, or unbiased reasons. Rules, and the words that form rules, are highly, even infinitely malleable, and have no objective force or meaning. Every case and every decision can go in any number of different ways, with no right or wrong, math-like answer. The argument is that objective truth itself is illusory, and that the law is a tool for reproducing hierarchy, not the rigorous, methodical search for a truth that doesn't exist.

The Precautionary Principle makes sense as a product of these modes of modern legal analysis. It would be sheer folly to hold meaningful environmental actions hostage to the illusory promise of eventual scientific advancements that would—someday—remove the troublesome unknowns from our equations. That day of certainty will never come. It can't be acceptable to allow something as omnipresent as uncertainty to serve as a universal hall-pass that forever excuses our collective rock-like lack of action on critical environmental issues. But how much uncertainty will justify our inaction, and under what specific circumstances?

The role of uncertainty in global-scale environmental decision-making can be especially troubling for attorneys, law professors, and law students—even though uncertainty is a familiar part of all law—because the system within which we must decide these awesome matters is broken. There's a hole in the middle of our system. It is missing something critically important, something universally understood as essential by everyone who works with the law. That missing piece is as simple yet as complex as this: Who has the burden of proof, and how steep is that burden?

Thanks to innumerable books, television programs, and movies in the genre of "courtroom drama," you don't have to be professionally involved with the law to comprehend the fundamental centrality of burden of proof to legal decision-making. The burden of proof is one of the most basic and most significant rules of the game. In many cases, the placement and magnitude of the burden of proof are outcome determinative, literally spelling the difference between victory and defeat for the parties involved. In American criminal trials, the government has the entire burden of proving that the defendant's behavior fits every necessary element (every absolutely essential, must-have building block) of the crime. The defendant in a criminal case has no burden of proof at all, and need not do anything to disprove any element of the crime he or she is charged with having committed. The defendant can merely sit back and watch, or take a long nap for that matter. If the court decides that the prosecution failed to prove even so much as one indispensable portion of the crime to the required degree of certainty, the defendant must be found not guilty of that crime.

And the amount of proof required of the prosecution is high—the famous proof beyond a reasonable doubt. That's a lot to ask of one side in any type of case, but because so much—a person's liberty and perhaps even life—hinges on the result of a criminal trial, our criminal justice system insists on this high level of confidence before allowing a defendant to be convicted. Because of the magnitude of the risk from an incorrect decision, i.e., the unjust deprivation of an innocent defendant's life or liberty, the system demands a sufficient quantity of legal and competent evidence to rule out any reasonable, rational alternative hypothesis other than the defendant's guilt. Unless we're crazy, we don't gamble a person's life on the flip of a coin. With that much at stake, the criminal justice system takes the stance that says "better safe than sorry" with a willingness to err on the side of innocence rather than guilt.

Famously, we would prefer to allow a large number of guilty defendants to go free rather than permit the unwarranted criminal conviction of even a single innocent accused person. In taking this stance, we implicitly accept the risk that some of the guilty people we acquit will use their freedom to commit additional crimes, including some extremely serious offenses such as rape and murder. We don't know what percentage of the acquitted guilty defendants will commit more crimes, nor what proportion of these crimes will be of the grievous variety, but we are prepared to accept those unknown but negative quantities as the price we pay for the system we value so greatly. We understand that we've set the bar high for our prosecutors, but that's the way we believe it should be within the confines of our free and just society.

Now let's contrast the burden of proof in criminal trials with the lower hurdle that applies in civil cases (such as torts/personal injury lawsuits, breach of contract actions, private property disputes, and actions brought under other non-criminal areas of the law). As with criminal cases, the party who initiates the civil legal proceedings (the plaintiff) has the burden of proving every element of the cause of action. The plaintiff is responsible for providing legally acceptable evidence sufficient to establish each necessary and indispensable element. If even one element is not adequately proved, the plaintiff loses. There's no grading on the curve and no partial credit; one missed element is as bad as missing all of them and has the same legal effect. The major distinction between civil and criminal trials lies with the amount of proof considered necessary to establish every element.

In civil trials, the plaintiff must prove each element by only a *preponderance of the evidence*, not beyond a reasonable doubt. Preponderance of the evidence is a tease of legalese much less familiar to those uncorrupted by formal legal education, but it is widely used across the panoramic spectrum of civil actions. It means, simply put, that every element—every essential piece—of the cause of action must be shown to be *more likely than not*. That's all. If the jury decides that any particular element is a true coin-flip with both alternatives equally probable (or that the defendant has the better of it), then the tie goes to the defendant and the plaintiff's case fails. Judgment for the defendant.

But if the plaintiff musters just the tiniest smidge of additional evidence, making it ever so slightly more than 50 percent likely that the element is met, everything changes. That microscopic extra bit of evidence placed gently upon the plaintiff's half of the finely-balanced scales of justice is all it takes to satisfy the missing element, and the tipping point is reached. Even a 50.000001 percent probability that the element was in the plaintiff's favor is sufficient to tip the scales decisively and win the case for the plaintiff. What a difference a smidge makes when we're playing under a preponderance of the evidence standard.

Of course, this minuscule incremental speck of tipping-point evidence is just a concept, and in reality we could never quantify evidence in this way. In a real courtroom, no one trots out a fancy-schmancy proof-o-meter with an easy-to-read light-up digital display that helpfully tells everyone how close the plaintiff has come to the promised land of 50 percent plus a smidge more. No one even knows for certain what probability figure actually describes the amount of proof before the court with any objectively quantitative measure at all, whether we're talking about 10 percent, 49 percent, 50.000001 percent, or 93 percent. The truth is that this is a subjective, qualitative, number-free decision we ask the jury to make for us. But the idea we give to the jury is the notion of preponderance of the evidence, the requirement that the plaintiff prove every element to the extent that they are at least a little bit more likely than not. A juror might visualize a scale and ask whether the plaintiff made it tip even a little to that side, or they might imagine our hypothetical digital read-out and look for anything so much as slightly beyond an even 50 percent, but the basic idea is the same.

Clearly, it's easier to prove a case under the preponderance of the evidence standard than it is when the requirement is proof beyond a reasonable doubt. Think of it this way. If you were taking my Constitutional Law class (my deepest sympathies), would you feel better about your chances of passing the course if you had to score 99% on the final exam to earn the credit, or if it only took 51%? Right! And you're not even a math major. The same reasoning applies in a courtroom. The party with the burden of proof will win significantly more cases under a preponderance standard, including some that he or she would have lost under the stricter beyond a reasonable doubt test. Our judicial system accepts this as a given, and as a desirable feature of civil actions.

But why? Why should plaintiffs have an easier time winning a personal injury lawsuit than a prosecutor in search of a criminal conviction?

The answer appears to relate to society's assessment of how much is at stake on both sides of the equation. Unlike criminal trials, civil cases lack the power to send anyone to prison, or worse. Money is generally what's at stake, not liberty or life itself. And in contrast to criminal cases, the government is not automatically on one side of the aisle in civil actions, putting all its considerable throw-weight, power, resources, and influence behind the push toward a particular outcome. Plus, we understand the value of enabling people—just ordinary private citizens in many instances—to gain redress for the legal harms they've suffered under a variety of circumstances. We see the value of making an injured party whole, to the extent a monetary award can compensate for what are often non-monetary losses like pain and suffering. In other words, society has an interest in giving plaintiffs a fair chance to win their cases and get back what someone made them lose. This is just another way we try to foster justice and fairness, similar to the way society wants the government to be able to convict and punish people who violate the law, so long as the government plays by the rules and can overcome the heavy presumption of innocence. So in civil actions, with legitimate and roughly equivalent interests on both sides of the balance, we have a see-saw standard that asks only a little more reason to rule for the plaintiff than for the defendant.

From a policy perspective, this is questionable. It's true that the big, bad leviathan government is not there with its giant thumb on one side of the scale to force a particular outcome in civil cases, and that the defendant's downside risk does not include any prospect of losing life or liberty, only money. These are considerable factors, but they are not the whole picture. The government can be a litigant in many civil trials, and in those cases it can still amass its tremendous resources to mount a formidable threat, aided by armies of attorneys and all the money it can print (or collect in taxes). In addition, civil cases can cause immense amounts of money to change hands, and the last time I checked, an avalanche of cash was a big deal. Depending on who wins and who loses, this seismic shift of greenbacks can bankrupt and ruin individuals or business associations, drive popular products out of the market, force major changes in the way people live and work, and many other formidable outcomes.

The aftershocks of a huge monetary award can spread far beyond the actual parties in a given case, dramatically affecting behavior of many others for good or for ill as observers adjust to either mimic or avoid similar results for themselves. This can happen in products liability cases where a design defect is found affecting a product or a complete family of related products, as well as in environmental litigation. Entire industries (such as asbestos) can be driven into extinction by ruinous civil liability, with wide-reaching effects on society, including unemployment, removal of products from the marketplace, extraction of previously installed products, development and marketing of substitutes, trading of one danger for another, and more.

Moreover, in civil actions where injunctive relief is sought, momentous changes in behavior can be driven by court-ordered injunctions to act or refrain from acting in a certain way. Injunctions have the potential to compel permanent, costly, difficult, and highly significant actions, or to prevent massive projects from being implemented. Such injunctions can affect large numbers of jobs and lives, arguably many more than would ever be impacted by any criminal case, even though they come from a civil and not a criminal court. Even between industries and between nations, injunctions can shift the competitive economic balance in favor of some and to the detriment of others. Yet the burden of proof remains the same preponderance of the evidence standard common to all garden-variety civil cases of much lesser potential importance.

With so much hanging on the particular burden of proof we select, and the outcome flipping from one extreme to the polar opposite in many cases, depending on the standard we decide to apply, it makes a huge difference whether any given case is treated as a criminal or as a civil matter. Some factual scenarios such as battery, assault, and trespass can fit into either category, and the choice of legal system can be outcome-determinative for not only the actual parties involved in the case but for the broader society as well. And given that this is true in the comparatively narrow and limited realm of judicial controversies, it is very much evident that the location and magnitude of the burden of proof are profoundly significant within the context of global environmental decision-making. Is the onus on those who want to impose new environmental controls, or on those whose activities allegedly threaten the environment? And how high is the hurdle over which the burdened party has to jump?

We've seen how the judicial system uses a very different burden of proof in the criminal versus the civil context, and that this probably relates at least partially to how much the parties and/or society have to lose from an incorrect decision. In the same way, we ought to ask whether the decisional burden of proof calculation we select for our most important environmental choices is context sensitive as well. In other words, does any of this vary depending on how much the proposed environmental measures would cost us in additional expenditures of money, work, quality of life, and foregone alternative opportunities, all things considered? Does any part of the burden of proof equation change based on the dimensions and/or likelihood of the feared environmental harm that could befall the planet if we fail to take effective corrective action? Do we take into account how much we stand to gain or lose in figuring out the way we should make our decisions? These threshold questions are at the heart of the struggle to do the right thing for ourselves and our world. Tragically, *we haven't answered them*. In a very real sense, we haven't really even acknowledged the questions. And that, in and of itself, is a major reason why we are stuck in the middle of all this risk and doubt, with no clear plan for how to get ourselves out. We're wandering through a minefield shrouded in thick fog, and we've never bothered to light our lantern. Got a light?

How, then, should we decide which way to bet with the planet at stake? What's the burden of proof when Earth is on trial, and on whose shoulders does its weight rest? How should decision-makers analyze momentous dilemmas such as the prospect of global warming, or a contemporary mass extinction, given the ineradicable doubts that will always hover around such subtle, widely dispersed, imperceptibly gradual phenomena? My books, *Ark of the Broken Covenant: Protecting the World's Biodiversity Hotspots*[8] and *Killing Our Oceans: Dealing with the Mass Extinction of Marine Life*[9] discuss a variation of Pascal's Wager I've developed for this type of situation. Within the specific context of the biodiversity crisis, I have called my analytical tool "the hotspots wager."

I originally named this decisional device the hotspots wager because I developed it and tailored it specifically to address the uncertainties associated with the potential for a modern mass extinction localized in the world's key biodiversity hotspots habitats. However, the same general method can be adapted to fit the issue of global climate change, transboundary air pollution, ozone layer depletion, energy policy, population pressures, or any other scenario in which we must make a tough environmental choice, with a lot at stake, without the benefit of air-tight underlying factual certitude on one or more relevant factors. Therefore, we can acknowledge this tool's fundamental flexibility and, in its more generalized form, give it the new name of "World Wager."

In choosing the term World Wager, I don't intend to skew the skewer in the direction of environmental activism as the default option. The name isn't supposed to be any type of suggestive message or leading question that implies, through the form in which it's phrased, the answer I want it to bring out. To be honest with you (and I can definitely be trusted, because after all, I am a lawyer) I like the name World Wager because it expends very few syllables, and it features the pleasing, if superficial, attraction of alliteration. But on a slightly less shallow level, the term World Wager also conveys a sense that we're talking about a situation of planet-wide concern, and not just a local or regional matter. Regardless of the outcome of our analysis, the World Wager tackles a worldwide phenomenon, with all the gravitas (and gravity) that any planet deserves. In other words, when the whole world is involved, there's a lot riding on the decision, so we should do all we can to get it right. As the old song goes (sort of), we've got the whole world in our hands, so let's not drop the ball by being careless.

By way of a concrete example of the World Wager, however, I will summarize the hotspot wager concept in its original form and refer to it by that more specific name. I want to use it to illustrate how we can integrate a reasoned and systematic approach to uncertainty into making strategic environmental decisions of the highest order. Afterwards, I'll suggest how we can aim a fundamentally similar approach at the climate change dilemma and other elusive targets.

The hotspots issue is one facet of a broader problem—a present-day mass extinction. One overarching decision needs to be made: Should the United States and other nations depart from the status quo and implement a scientifically sound program of protection and preservation of the world's biodiversity hotspots? An effective hotspots program could be a vital piece in the puzzle of stopping the contemporary mass extinction, but there are three substantial unknowns in the equation, lurking as landmines to dissuade wary leaders from taking action. For politicians looking for any arm-waving excuse to stand pat, even one questionable factor often suffices to call for the ever-popular and literally effortless do-nothing option. The presence of three unknowns virtually guarantees inaction. But is that the correct way to handle the situation, with so much at stake? Let's look a little closer.

The three chief unknowns in the hotspots wager are as follows. (1) How many species (including species no one has ever discovered) now are endemic, that is, entirely confined, to the world's biodiversity hotspots? (2) What is the probability that many or most of those hotspots-endemic species (known or unknown) will become extinct within the foreseeable future if no one takes action to save them? (3) What is the tangible practical value of those endangered, hotspots-endemic species to human beings, now or in the future?

Without a paradigm-shift, many political leaders and their constituents take a quick glance at the issue (at best) and decide that the combination of these three unknowns provides more than enough justification to forget the whole thing and turn their attention elsewhere. Money could be spent on other projects, and effort could be expended on near-term, nearby causes where everyone can easily see a real and immediate need for intervention. From this home-grown, home-turf perspective, the trio of big questions moves the hotspots issue from the front burner all the way back to Fantasyland. Choose your favorite cliché: We have bigger fish to fry; we've got better things to do; we have other things to worry about; it's the Chicken Little environmentalists crying "the sky is falling" again; we need to put our own house in order first. This cavalry of clichés has charged in on the backs of the question marks and routed many attempts to secure meaningful protection for various individual endangered wildlife habitats, both on land and in the oceans. Moreover, no political leader anywhere has ever even tried to marshal support for any comprehensive legal safeguards for the world's most important priority-ranked habitats. On a global, systemic level, the hotspots crisis has never made a single blip on the political-legal radar screen. The unknowns have undone the mass extinction issue to an even greater extent than with climate change.

The hotspots wager asks people to frame the issue differently. As with Pascal's Wager, the hotspots wager requires, as the first step, an analysis of the reasonably likely answers to each of the three questions. To simplify the matter enormously and facilitate clear decision-making, we can limit this to two polar opposite answers, one high and one low, for each unknown. Then, for each extreme value of each question, the hotspots wager calls for determining the probable outcomes that might flow out of our decision either to do nothing further to protect the hotspots or to implement systematic protections. This inquiry will produce best-case and worst-case results for each unknown, depending on the true (but unknowable) value the unknown actually has.

Because there are three unknowns relevant to the hotspots wager, we need to incorporate a means of combining the polar opposite values of all three unknowns simultaneously. We can do this graphically, or in tabular form, to represent what happens when the true values of all three unknowns are combined and compared with all the others. In its original form I called the table of these permutations the "hotspots decision matrix."[10] Again, if we are using a variation of this approach to help us resolve other planet-wide environmental problems such as climate change, the basic method would still apply. So in recognition of its broader application, I have renamed the more generic form the "world decision matrix." Once we've listed every possible permutation of the polar values of all three unknowns combined, we can assess the best- and worst-case consequences for each permutation. This again depends on whether or not we decide to institute new, dramatically enhanced measures to protect the hotspots. I will leave a detailed analysis of that question until Chapter 10, after I've established some additional foundation to help us make a reasoned decision. For now, let me propose one possible non-crazy way in which we might fill in the blanks in our world decision matrix.

The World Wager in general, or the hotspots wager in particular, doesn't force us to limit ourselves to only two polar opposite values for each unknown. It is realistic to presume that the actual answer for any given unknown could be literally anywhere between the extreme maximum and minimum values. For purposes of hotspots wager calculations, we are free to select as many different values as we'd like for all three unknowns.[11] We could postulate a range of possible values for each unknown, and then derive all permutations of these. Of course, by choosing more than two possible values for each unknown, we greatly multiply the number of available permutations. Any table or pie-chart based on such refined analysis would have to be much larger and more complex. Still, the core principles would remain the same. Whatever intermediate value we imagine for any of the three unknowns, when we combine it with any value for the other two unknowns, we can then gauge the magnitude of the resulting gain or loss from a decision to protect or not to protect the hotspots. It becomes more unwieldy, but it is still entirely feasible.

We could also incorporate another realistic refinement by adjusting for our best estimation of the relative probability of every proposed value of each unknown. That is, based on the best available evidence, we might conclude that it is more probable that numerous species are endemic to the hotspots, including many that haven't yet been discovered, than the alternative (that there are no additional species still waiting to be identified in the hotspots). Similarly, given what we know about the rapid destruction of hotspots habitat and the effect habitat loss typically exerts on extinction rates, we could postulate a higher probability for a serious extinction threat than for a low risk of extinction in the hotspots. We might also reasonably propose that, given the richness of hotspots biodiversity and the almost total absence of intensive scientific study of that biota up to now, some species are likely to be endemic to the hotspots that one day could be the basis for the new penicillin, holding the key to saving untold millions of human lives.[12] Based on our assessment of the relative likelihood of each possible value for each of the three unknowns, we could assign a best-guess probability figure for every one of the permutations, and thus for the results associated with them. If we converted the decision matrix into a pie-chart or other graphical representation, then, some slices of the pie would be larger than others. Indeed, there is absolutely no reason to expect that every one of the possible permutations is equally probable. But again, for sake of simplicity, we can omit any probability estimates from the decision matrix. Let's make it easy on ourselves, as usual.

In its most basic form, we can reasonably suggest that the hotspots wager yields a decision matrix in the following form.

ENACT AND FUND MAJOR HOTSPOTS PROTECTION?	TRUE DEGREE OF EXTINCTION RISK IN HOTSPOTS	TRUE NUMBER OF UNKNOWN SPECIES IN HOTSPOTS	TRUE TANGIBLE VALUE OF ALL SPECIES IN HOTSPOTS	RESULTS OF HOTSPOTS FUNDING DECISION
NO	LOW	LOW	LOW	LUCKY WAGER, MONEY SAVED
NO	HIGH	LOW	LOW	SECOND ORDER SERIOUS ERROR
NO	LOW	HIGH	LOW	LUCKY WAGER, MONEY SAVED
NO	HIGH	HIGH	LOW	FIRST ORDER SERIOUS ERROR
NO	LOW	LOW	HIGH	LUCKY WAGER, MONEY SAVED
NO	HIGH	LOW	HIGH	SECOND ORDER GRAVE ERROR
NO	LOW	HIGH	HIGH	LUCKY WAGER, MONEY SAVED
NO	HIGH	HIGH	HIGH	FIRST ORDER GRAVE ERROR
YES	LOW	LOW	LOW	UNUSED INSURANCE
YES	HIGH	LOW	LOW	SECOND ORDER SOFT BENEFIT
YES	LOW	HIGH	LOW	UNUSED INSURANCE
YES	HIGH	HIGH	LOW	FIRST ORDER SOFT BENEFIT
YES	LOW	LOW	HIGH	UNUSED INSURANCE
YES	HIGH	LOW	HIGH	SECOND ORDER JACKPOT
YES	LOW	HIGH	HIGH	UNUSED INSURANCE
YES	HIGH	HIGH	HIGH	FIRST ORDER JACKPOT

The results column reflects my attempt to describe, qualitatively if not quantitatively, the dimension—the size and nature—of the positive or negative results from every possible outcome. A full discussion is included in some of my previous works, and to some extent later in this book's Chapter 10. For conceptual purposes, it suffices to assign some recognizable description to the magnitude and persistence of every possible result, good or bad, to allow us to make comparisons of relative impacts. To make a wise choice, we need to know which decisions might produce great news, or really awful news, or everything in between. But let's put the hotspots on the back burner for a moment now and see how we might use the same analytical tools to turn up the heat on the global warming issue.

How can the World Wager be altered to apply to the climate change decision as well as the hotspots? Again, it is uttely crucial to choose well when selecting the primary variables and when assessing the range and significance of their possible values. This requires a very frank and honest, open approach to all the available evidence so as not to bias the process toward a predetermined conclusion. Like any tool we leave within reach of humans, the World Wager can be misused if the people who wield it are corrupt, ignorant, misguided, irrational, or determined to disregard the warning label. The ancient truth "garbage in, garbage out" is fully applicable to the World Wager, so it behooves us to keep our input as free of garbage as we can. Remember the butterfly effect and how the slightest difference in ingredients can spell a huge difference in how it all turns out.

To minimize the chance that our output will be indistinguishable from garbage, we need to pick the right variables for the "climate wager" form of the World Wager, with accommodation for uncertainty in the underlying evidence. This exercise can be repugnant to some intensely interested parties on both sides of the climate change chasm; they've dug themselves into deep, heavily reinforced trenches in their battle lines suitable for any full-scale live-ammo reenactment of Verdun. For true believers in climate change, and for their polar opposites (true skeptics? fervently confirmed doubters?) there is *no legitimate other side* to the debate...and so there really is no debate, just a soliloquy or a monologue. But it's hard to win new converts when you're always just preaching to the choir already holding reserved seats inside the church at a private service.

I like dogma as much as the next guy (though I prefer the dog as man's best friend), but any dogma worth its devotees and its dawgs should be sufficiently robust to withstand an evenhanded airing of the heretics' supposed scandalously false doctrines. Seriously, if the opposition is completely off-base, there should be no harm in letting everyone hear their deep-down stupid arguments and made-up fairy tale fictions. The marketplace of ideas depends on free and open competition of all viewpoints, from great to gross, so that people can make the best choices for themselves with full awareness of all the products available.

In fact, the World Wager is specifically designed to expose the relative strengths and vulnerabilities in any problem and equip rational people with the means to make correct, well-informed decisions in light of all the relevant evidence. If some notions are so half-baked that they can't stand this heat, it's worth our trouble to prove it. So let's select a reasonable set of variables for inclusion in the climate wager with this in mind, and give the tool a chance to do its work.

Here are my candidates for a credible, and bare-minimum-short list of essential factors tied to the climate wager, with acknowledgement that other, somewhat different selections could also be appropriate: (1) The extent to which Earth's overall temperature is now rising at a rate greater than the acceptable historical norm and will continue to increase above its present level in the absence of effective human intervention; (2) The quantum of harm, on balance, that would be caused by a further climb in the planet's temperature of a magnitude similar to that predicted in factor 1; and (3) The extent to which humans collectively are able to make significant, timely, and adequately large modifications of Earth's mean temperature in a selectively targeted direction and amount.

When the climate wager is framed in this way, it includes three variables, just as there are three featured in the hotspots wager. Clearly, it would be defensible to add one or two more to the number of unknowns in the climate wager, as I'll explain shortly. But an increase in the number of variables greatly multiplies the complexity of any problem, making it more difficult to solve. In deference to the unimaginably high stakes riding on our decision here, I made every effort to choose the absolute minimum number of variables acceptable for the climate wager. In deliberately opting for an abbreviated list of unknowns, I purposefully chose to err on the side of favoring the environmentally-proactive alternative. But did I go far enough in this direction?

Could we justifiably omit any of the three short-list variables without inflicting an unacceptable level of damage on the decision-making process? Not in my opinion, unfortunately. I consider each of the three unknowns as a necessary, indispensable cog in the mechanism. Each one of the variables focuses on an important aspect of the climate change challenge that is not adequately addressed by either of the others. And although a degree of interrelatedness exists between and among the variables, none is actually just a portion of any other; each expresses something unique and important not otherwise embraced. In fact, I considered adding *at least* one more variable rather than deleting any of the three, and there would be considerable justification for doing so.

For example, we could add a fourth factor touching on the length of time that will (or might) elapse before major damage results from climate change. This could even, with some support, be accompanied by a related but separate fifth factor dealing with the length of time remaining until it is *too late* to avert any significant and widespread climate-related damage. I decided to err on the side of introducing fewer unknowns, and did not include either of these variables, although either or both of them would contribute some acknowledgement of the pivotal elements of imminence and irreversible harm. They each would reveal vital information relating to urgency, but in two somewhat different respects. Am I on solid ground (or at least solidly frozen ice) in making a judgment call that excludes both unknowns?

As we decide what if anything to do about the planet's temperature, it's important that we know both how much time we have before serious harm would be actually caused *and* how long we can afford to wait before implementing any remediation measures without guaranteeing dire consequences. If either or both of those numbers are low, that could be a major force in favor of swiftly mobilizing to combat climate change, while a high value might influence us to wait before expending enormous resources on a non-emergent problem (and give ourselves time to work on developing less disruptive and more affordable corrective measures). The luxury of more time would also let us take a wait-and-see approach and gather more information, thereby reducing the level of uncertainty, before we commit to a vast and expensive program. The values of those two imminence unknowns could also drive our course of action in terms of prevention or mitigation of harms already set in motion and already inevitable, as well as proactive preventive measures designed to avert other consequences that are still avoidable.

All of this is worthwhile and important. Still, I decided that factor 2 standing alone can serve as a reasonably good surrogate for the imminence issue, and so I left out these other candidates. In my view, it is acceptable to rely on this single unknown to capture the relevant information on both the magnitude and the imminence of the harm we would suffer in the absence of effective human intervention on climate change. But because I opted to omit any expressly time-related unknowns from the climate wager, it is very important that we look at the second variable comprehensively. Therefore, our appraisal of variable 2 should incorporate an assessment of *how soon* we are likely to suffer highly impactful or irreversible levels of climate-sparked harm, as well as how much time we still have to work with *until such harms become inevitable*. Because of my conscious decision to employ the absolute minimum number of unknowns in the climate wager, we'll ask variable 2 to do some heavy lifting. It must bear the weight of how much damage we will suffer if we remain on our present course *and* how much time we have left until it's too late to steer clear of the melting icebergs in the rising waters ahead.

As I mentioned in Chapter 2, the "Climategate" scandal that broke in late 2009 calls into question some of the most significant sources of data on the dimensions and nature of the underlying climate change threat, including the legitimacy of the "hockey stick graph" and the recent trends in planetary average temperatures.[13] By rendering suspect the validity of supposedly objective evidence, this controversy has thrown another helping of doubt onto our disorderly heap of factors and questions. In light of this, variable 2 becomes even more crucial in terms of the evidence we consider and the evidence we exclude. It is of paramount importance that we be as scrupulously evenhanded and open-minded as possible in assessing the relative merits and weaknesses of all the relevant data that should properly be factored into this variable. If we allow ourselves to succumb to the same type of biases and foreordained conclusions that undermined the integrity of the people at the Climatic Research Unit, we will rob the climate wager of its power to lead us to fair, rational, evidence-based, objective decisions. We will instead get only the answer we wanted, and the only answer made possible by our own "garbage-in" contortion, manipulation, and restriction of the variables. And what's the point of that? It isn't really a decision if we artificially remove all possibilities but one in advance. It's just a foregone conclusion dressed up in a disguise of false fairness.

You, or any decision-maker, are entirely free to modify my version of the climate wager to add the excluded time-sensitive factors, or others, or even to remove any variables. The chief concern in doing so is that every variable must be selected with meticulous care. If we add new factors or cut existing ones without sufficient justification, we could nudge the results in a misleading direction and unleash a stampede of chaotic butterflies. Well maybe that doesn't sound scary enough, but you see what I mean. It is very important that we be scrupulously evenhanded and open-minded on the threshold question of which unknowns to include. This is no place for gerrymandering, butterfly ballots, hanging chads, or vote fraud. In light of the fact that I've already trimmed down the number of variables to just a bone-core three, it is especially crucial to exercise the utmost caution in making further reductions. In fact, we could make a much stronger case for going in the opposite direction and incorporating additional variables.

I've already mentioned that I chose to omit at least one and possibly two variables related to *how soon* we are likely to see sizable climate-driven damage and/or how long the fuse might continue to burn until it's *too late* to prevent devastating big-bang damage. There are other obvious opportunities to add more unknowns to the climate wager, which I also decided against in a desperate attempt to keep the decision matrix within marginally manageable levels of complexity. For example, I could have supplemented the list with a factor aimed at the extent to which any current rise in mean planetary temperature is attributable to solar events or other phenomena not driven by *human activity*. Such information could tend to show that people are either more responsible or less responsible than natural causes for the modern climatic trend. This unknown might also demonstrate that we have either significant or minimal ability to change the direction of that trend, regardless of its cause. I opted to rule out this factor because I felt that its primary thrust was adequately covered in some form by the existing list. Also, even if harmful climate change is not primarily anthropogenic, we may well want to do everything we can to dodge the bullet anyway, just because we don't like being hit by bullets whether or not we're innocent.

Likewise, I might have added a factor looking to the *possible benefits* from higher planetary temperatures. These could include the prospect of lengthier growing seasons for crops and other plants; added farm- and graze-ready available land (due to receding ice and snow cover); better conditions for photosynthesizing plants to flourish generally; enhanced rates of removal of carbon dioxide from the atmosphere through revved-up carbon fixation from all those bigger and more plentiful plants; and other theoretically salutary effects. Again, I decided not to make this rather contrarian notion a separate factor, because I viewed these potentially counterbalancing positive effects as being already fairly embraced within the overall calculation of factor 2. This will be true, however, only if we examine that factor fully and equitably as a focus on the *net* harm caused on balance by unchecked climate change.

Just as we did with the hotspots, we also must determine, to the best of our limited abilities, the range of consequences possible from every combination of the three climate change variables. All of this is viewed from the perspective of a decision whether or not to take a significant amount of action reasonably likely to mitigate the feared harm that could be caused by unopposed global climate alteration. To simplify, we'll again consider this decision as an all-or-nothing, yes or no matter, although clearly many intermediate courses of action are available, in dizzying variety, each with its own prospects of success and its own costs and risks. Similarly, we will speculate as to the absolute maximum and minimum polar-extreme actual values that could conceivably exist in reality for every one of the variables. We'll do this even though we acknowledge that such certain and precise information on the underlying facts is now and may forever be beyond our grasp; we go through the exercise to illuminate what it *could* mean to us if we could somehow gain this perfect omniscience. And, once more, we'll follow the same procedure we used with the hotspots and focus on just the extreme best-case and worst-case outcomes that could possibly result from every permutation of values for the unknowns and action decisions.

Not to continue beating a horse that's already as dead as a horse can be, but the climate wager doesn't necessarily require us to limit our analysis to these polar extremes in any respect. We could still use the wager if we allowed for the possibility of compromise levels of intervention, and/or intermediate "true facts" on the hidden value of each variable. We could also incorporate a probability estimate gauging the likelihood of the various outcomes that might flow from every marriage of choices and factor values. This more nuanced and realistic approach wouldn't disturb the core principles underlying the wager and decision matrix. The same general precepts and tendencies would still apply. But the analysis would be much more lengthy and convoluted, and the matrix would become unwieldy and excessively large as it expanded to accommodate all those extra gradations and subtleties. More important, it would be *way* too much work, and we all badly need a cheesecake break. Trust me. We can still see the big picture while keeping the details as simple as possible, with the understanding that plenty of room abounds for fine brush strokes within the surrounding framework.

Here, then, is just one reasonable way of expressing the climate wager in its decision matrix form. As I've explained, wherever possible I chose to minimize the number of unknowns and thus frame this issue in the mode most favorable and conducive to an environmentally-protective and proactive paradigm.

LAUNCH MAJOR GREENHOUSE GAS CONTROL WORLDWIDE CAMPAIGN?	TRUE AMOUNT OF FURTHER ABNORMAL TEMPERATURE RISE IN FUTURE IF LEFT UNCHECKED	TRUE EXTENT OF NET HARM FROM FUTURE TEMPERATURE RISE IF LEFT UNCHECKED	TRUE DEGREE OF HUMAN ABILITY TO SHIFT CLIMATE ADEQUATELY BY DESIGN	RESULT OF DECISION ON GREENHOUSE GAS CONTROL CAMPAIGN
NO	LOW	LOW	LOW	LUCKY WAGER, MONEY AND OPPORTUNITIES SAVED
NO	HIGH	LOW	LOW	LUCKY WAGER, MONEY AND OPPORTUNITIES SAVED
NO	LOW	HIGH	LOW	LUCKY WAGER, MONEY AND OPPORTUNITIES SAVED
NO	HIGH	HIGH	LOW	LUCKY WAGER, MONEY AND OPPORTUNITIES SAVED
NO	LOW	LOW	HIGH	LUCKY WAGER, MONEY AND OPPORTUNITIES SAVED
NO	HIGH	LOW	HIGH	LUCKY WAGER, MONEY AND OPPORTUNITIES SAVED
NO	LOW	HIGH	HIGH	FIRST ORDER GRAVE ERROR
NO	HIGH	HIGH	HIGH	FIRST ORDER GRAVE ERROR
YES	LOW	LOW	LOW	UNUSED AND VERY COSTLY INSURANCE
YES	HIGH	LOW	LOW	UNUSED AND VERY COSTLY INSURANCE
YES	LOW	HIGH	LOW	UNUSED AND VERY COSTLY INSURANCE
YES	HIGH	HIGH	LOW	UNUSED AND VERY COSTLY INSURANCE
YES	LOW	LOW	HIGH	UNUSED AND VERY COSTLY INSURANCE
YES	HIGH	LOW	HIGH	UNUSED AND VERY COSTLY INSURANCE
YES	LOW	HIGH	HIGH	FIRST ORDER JACKPOT
YES	HIGH	HIGH	HIGH	FIRST ORDER JACKPOT

Once again, I will mostly defer in-depth analysis of this wager until Chapter 10. However, because this decision matrix has never before appeared in print anywhere, I want to highlight some key features. For example, the climate version of the world decision matrix raises and emphasizes a previously neglected point relevant to the Precautionary Principle and related analysis. That point is that all equations involving uncertainty are not created equal. Just as the idiosyncrasies of each variant of uncertainty are important to a determination of whether corrective action is prudent, the cost and added burden of an effective response is also a major factor.

For example, the hotspots wager might seem to point toward the need for legal action to forestall a mass extinction, *but* this also depends on the magnitude of the expense and inconvenience involved in mounting a countermeasure that would adequately get the job done, assuming such a response is even feasible. We need to know not only that *some* legal corrective action is called for, but also *what level* of action and what it would *cost*, in dollars and disruption. The same is true to an even greater extent for the climate wager. To craft a wise decision, we would need to account for all significant and reasonably probable attributes of our investment in a response to the perceived climate-change threat. A balanced, objective assessment should fully consider the likely costs and benefits, credits and debits, rewards and risks associated with the proposed action. This evaluation of the downside potential of our climate intervention campaign would look at all the costs, in both size and likelihood. These costs might include the amount of our commitment in terms of money, material, person-hours, lost opportunity costs, higher energy prices, effort, inconvenience, disruption to normal activities and quality of life, reduction in economic productivity and ability to compete, trade barriers, increased unemployment, undesirable side-effects, region-specific hardships, and others.

To the extent we can estimate and aggregate such investment

factors to a satisfactory degree, we should balance them against the magnitude and probability of the threat, as derived from our analysis along the lines of the hotspots wager. The two variables can interact in a variety of ways along a continuum of possibilities. A very large and reasonably probable threat could rationally justify quite a formidable mobilization in opposition, pulling out all the stops in the proverbial "moral equivalent of war." Conversely, if an effective response would be relatively easy and inexpensive, we should require less evidence of a likely disaster before we conclude that we need to take that type of action. It's analogous to the way we set different burdens of proof in civil and criminal cases according to how much is at risk. This may seem an obvious point, but in an uncertain universe desperate problems call for desperate measures, while more pedestrian threats can be properly met with more modest resources or even ignored.

The climate wager and decision matrix implicate other central issues as well. Notably, there is also an element of uncertainty as to whether any particular effort to correct an environmental problem will in fact work at all, or to what extent it will be effective in serving the intended purpose. There are always unknowns inherent in any human endeavor. No matter how good our intentions, no matter how much our efforts cost us and how much we sacrifice, and no matter how careful our planning and theorizing, the actual results are always hostage to the vagaries of probability and uncertainty. We can spend vast sums of treasure, time, and toil on a full-scale attempt to solve a problem, but that is no guarantee that our Herculean effort will be successful, or even that it will do more good than harm. Murphy's Law teaches us that we could try mightily, like Hercules, to clean out those seriously nasty Augean Stables, and succeed only in spreading the mess around and making the organic waste matter flow downhill toward our beach house. Anyone who has ever launched a self-help home improvement project, or called a plumber or electrician to rush to the rescue of a domestic emergency, can attest to the validity of this immutable principle.

For these reasons, the decision matrix within the climate context contains some different terms in the results column from those we saw in the hotspots version. To a far greater extent than with hotspots preservation, any campaign to reduce significantly the Earth's overall temperature would necessarily entail enormous changes in the way billions of people live and work. There would be huge opportunity costs as entire industries vanish or shrink and probably major adjustments in quality and method of life. Climate intervention would require much more than the few billion dollars annually needed to protect and maintain the relatively small territory collectively occupied by all the world's hotspots and to make up for the lost opportunities to farm, graze, hunt, reside, or cut timber in them. Thus, when we expend massive resources and engineer fundamental reinvention of large components of human industry and society, if this turns out to have been unnecessary or ineffective, we have made a very costly mistake indeed.

We might still call this "unused insurance" as we did with the hotspots matrix, but that's more than a little misleading. The level of expenditure, effort, and restructuring mandated for a credible climate change campaign is so immense that, if needless or useless, it could only be thought of as unused insurance of the costliest and most burdensome variety. It would not be analogous to the moderately-priced life, health, or home insurance premiums many of us take in stride as a prudent and reasonably affordable investment in protecting our families and our way of life. If it's similar to any form of insurance, it might be the extremely expensive medical malpractice insurance used by some physicians, especially those with practices that entail a high risk of potentially ruinous multi-million-dollar malpractice lawsuits. That variety of insurance can be prohibitively expensive and onerous, to the extent that some doctors stop handling the riskiest types of cases, or move to states where the tort system is less favorable to plaintiffs. It's not just a routine cost-of-living or cost-of-doing-business type of expense most of us take in stride every year.

Similarly, if we decide not to wage a gigantic, multi-pronged campaign to combat global climate change, and if it turns out that such a campaign either was not needed or wouldn't have made enough of a difference anyway, we have saved more than "just" a few billion dollars and moderate disruption every year. For better or for worse, under those circumstances our inaction would be rewarded with preservation of the fundamental economic and social structures and mechanisms that would have been dramatically transformed by a climate counteroffensive. These systemic reinventions are of a greater order of magnitude than any adjustments implicated by hotspots conservation, and so the results associated with them need to reflect *how much* we are "wagering" when we decide to push all those chips into the middle of the table. When we make a decision not to go there, and events providentially prove our decision correct, that's a tremendously significant savings, both in terms of cash expenditures and opportunity costs.

As much as we might long for metaphysical certitude in our lives, we can still comprehend an inevitable element of uncertainty in most if not all situations. We may not know anything about the properties of subatomic particles or axiomatic mathematical systems, but we intuitively know how much random chance, rather than bedrock, undeniably true dogma dominates our daily experience. And if we accept the necessity of incorporating probabilistic, odds-related, elusive variables into our personal major life decisions (such as whether to insure our home or our life), we could be open to an argument by analogy that the same should apply in environmental matters that could drastically affect all life on Earth.

But it may be even more complex than that, and an even greater challenge to our typical way of assessing risks, benefits, probabilities, and the relative value of insurance. There are features of global environmental threats that make the insurance analogy an imperfect fit, at least from some perspectives. For example, very slow-moving dangers such as mass extinction (and, to a lesser extent, climate change) may take hundreds of years—or even more—to reach the point of no return, and even after reaching that point, the effects may not be obvious for far longer. If key species gradually slip into oblivion, this will likely be invisible to most people, and the ecological services or other benefits those species provided may not be palpably diminished for a very long time. This extreme gradualism and long-term horizon is very different from the hazards against which people typically insure. When we buy insurance for our life, our health, or our home, we weigh the prospects of needing that insurance during the next year, or perhaps the next few years. We don't buy insurance for the next thousand years.

Clearly, in our own lives, we have no need to concern ourselves over the probability of needing life or health insurance beyond our own lifetime. Whether we think we'll live another 5 years or another 50, that span is real and meaningful to us personally, in a way that a span of hundreds or thousands of years is not. And the people who would benefit directly from our life insurance or property insurance are people we know personally—usually our relatives or our close friends, and maybe their children or grandchildren. They are not remote descendants or unrelated members of the greatly-extended human family dozens of generations removed from us. *Do we even care* whether actions we take today might save people a thousand years from now? Would the chance of averting a disastrous pandemic or ecological catastrophe for people who might be alive three thousand years in the future be of any concern to us now, as we weigh whether to invest in long-term environmental insurance?

In addition, perils feel more genuine to us if we have actually experienced something akin to them in our own lives, or at least if someone we know has been hit by them. Many of us have some familiarity with destructive fires, earthquakes, floods, serious health problems, and other dark clouds for which we might take out insurance. But would we recognize minuscule trends in climate, or gradual loss of vital ecosystems, or extremely drawn-out decimation of endangered species in some remote forest? We can see and feel the impact of menaces to our immediate personal life, health, and home. But many of us never even try to imagine what it would look like if worldwide climate shifts or biodiversity were inching ever so slowly toward a fog-enshrouded far-off cliff. It doesn't hit us where we live, and it doesn't hit us where it hurts. At least it doesn't feel like it, and so we don't pay attention the same way we do when natural disasters or personal tragedies jeopardize our own family or our next-door neighbors.

When the issue isn't protecting the house we now live in, or our own financial security, or our immediate family's welfare, but rather insuring people we'll never know against a possible harm long past our lifetime, how does our decision-making differ? We'd still need to pay the premiums now, ourselves, in cooperation with millions of other taxpayers, to safeguard important habitats. The expense, sacrifices, and inconveniences would be ours, just as when we deny ourselves other things to pay for our life insurance every year. But the premiums to protect the world's biodiversity hotspots would have to be paid every year...forever. The life we're insuring wouldn't be our own all-too-short life, but instead the life of our species and all the other species on this planet. Now *that's* a long-term commitment. Are we prepared to pay, and to commit our descendants to pay, premiums on preserving our planet's environment indefinitely, as a cost of living on Earth, without any end point? That's a lot of money, invested year after year for all time. The harm it might avert could still vastly outweigh the amount we spend to prevent it, but the investment is both tremendously prolonged and large, nonetheless.

And who would benefit if we turn out, one day way off in

Tomorrowland, to stave off a great evil? Our anonymous, unborn beneficiaries might include some of our great-great-many-great-grandchildren, whose pictures we'll never carry in our wallet. But they will also include many more persons who will live in houses we've never visited, who adhere to religions we don't share, who live lifestyles we find repugnant, and whose politics we think misguided. Come to think of it, that might also describe a lot of our own children. But you see my point.

Will we buy insurance to protect these far-removed people long into the centuries beyond our own from dangers we didn't personally cause and that probably will never threaten us directly? Here we aren't talking about the chance of a fire destroying our own home, or cancer devastating our health, or our family left homeless by our untimely death. No, the specter of a creeping, insidious mass extinction or other imperceptibly menacing environmental risk is so hidden, slow, widespread, and long-deferred that it may be entirely off our radar screen. We have no experience paying for insurance that protects the whole planet, including people in places on the other side of the globe who will live many centuries from now. Other than perhaps fighting in world wars, where much of the good and evil directly affects nations and persons far away, we live our lives primarily in the near-term and in the near-distance. Can we broaden our horizon to encompass precautions against a planet-wide disaster that may be stealthily sneaking up on us, in places we've never seen, over the course of dozens of lifetimes? Is it something we should insure against? That's a decision quite distinct from the one we face when we look into our own children's eyes and ask ourselves whether we should spend the money it takes to buy life insurance to protect them.

It is not the type of question we have much experience even pondering. Do most of us spend any time contemplating our lasting legacy, or the planet's future many centuries after we're gone? We usually have more than enough to worry about within the timeframe of the coming week, let alone millennia beyond our lifetime. We will very likely never meet or know the names of the people we might help by taking environmental precautions today against hyper-gradual threats like mass extinction or even relatively near-term hazards like climate change. Their names will never be written as beneficiaries on any insurance policy we take out. So does that mean we don't care whether they live or die? Does that mean it's all the same to us whether they have a world that's livable, or worth living in? Or does their remoteness from us in time and place shove them forever beyond our concern as people we think about as we decide how we live our lives? This is a crucial question, and a key part of the gamble we're taking right now as we're betting the Earth. And it's a question many of us have never even considered. If, after us, comes the deluge—and if it will probably come, if at all, centuries after we're gone—do we care?

I don't know how most people would feel about this. Does it matter to us if the human race dies out a few hundred years from now? Do we care if there are people around on Earth two thousand years after our grandkids are dead? Is it worth our time, money, and trouble to take action now to make this planet livable for people and creatures that might walk the Earth dozens of centuries after we're nothing but dust? I've never conducted a survey to ask these vital questions. *But I will ask you.* What do you think? Do these things matter to you, personally? How far down the road do you care whether there is life on Earth or a world that's fit for people? How far into the future do you stop feeling a link to whether the human race will still survive? How many centuries past tomorrow do you cease to have a stake in the outcome of whether people will continue to live, laugh, and love? Is 100 years the limit of your vision? Or 200 years? Or 2,000 years? Think about it. Think about how you view *your connection* to the people who might follow you on this planet and whether they are worth the insurance we could buy for them today. The answer is absolutely at the heart of this global game of chance we're playing right now.

With this basic understanding of the World Wager and decision matrix as our jumping-off point, I want to devote the next two chapters to matters a little easier for most of us to appreciate. I'll look at examples we deal with in our own lives that in some ways have features in common with the global issues of biodiversity loss and climate change. If we can see how we all manage to manage uncertainty, incomplete information, probabilities, risk, costs, and benefits as we make many of our personal decisions, we'll be better prepared to apply those same basic principles to much grander dimensions. So let's go take some chances!

CHAPTER 8

You Bet Your Life, and Here's How

It's an insidious and deadly trap that uncertainty sets for us. Some of the greatest dangers the world has ever known are advancing unchecked because people who might take action choose to wait. And what are they waiting for? They see the unanswered questions, the variables without defined values, and the gaps in our evidence. The one thing they see with certainty is the uncertainty. And so they follow what feels to them like a reasonable course under the unsettled circumstances. They stay put. They delay. They insist on postponing any movement until the scientific evidence stacks up high enough to satisfy what they might imagine to be the burden of proof.

Yet this mindset is as misguided as it is prevalent. It's a perverse reversal of the Precautionary Principle. It demands an unrealistic or even impossible level of scientific certainty as a prerequisite to corrective action in matters so monumental that the wrong decision could usher in a global catastrophe. It fails to recognize that deciding to postpone intervention is the same as deciding not to intervene at all in situations demanding a swift proactive push. This negative-image corruption of the Precautionary Principle mires us in catatonia in the face of cataclysm.

The Precautionary Principle is sometimes derided as a vehicle for concocting a "guilty until proven innocent" prisoner's dilemma.

My proposal, however, should provide an effective safeguard against such kangaroo-court abuses. By acknowledging the ineluctable uncertainty that cloaks our choices, we can avoid the opposite mistake—refusing to act whenever we find any scientific dispute—while also ensuring that the deck isn't stacked in favor of knee-jerk (over)-reaction to every putative scare. Pascal, Gödel, and Heisenberg infuse my World Wager and its many possible variants with a reasonable and realistic means of incorporating unknowns into rational decision-making. The result doesn't presume we should always charge into crisis mode whenever we see a glimpse of a red flag. On the contrary, this approach aims to stop us from jamming our collective foot on the brake every time we think of an unanswered question. I know it's a radical idea, but my method asks us to take a little time to think before we act…or before we fail to act. Once we think things through, we can then make a reasonable choice on which pedal to step on—the accelerator *or* the brake.

Proper application of the World Wager makes room for a reasoned estimate of the relative probabilities of each relevant possibility. A remote chance of some disaster should register on our radar screen, and we need to consider all available parameters before deciding for or against taking action. Many times, a potential hazard is so improbable, and the costs of eliminating or significantly reducing its already-minuscule chance of occurring are so immense, that the prudent outcome is the no-action alternative. Neither in our personal lives nor in our large-scale environmental policy can we afford to cower in persistent terror of every possible danger, however unlikely. That attitude brings fearful paralysis of useful activity and improvident, wasteful overprotection against unrealistic bogeymen. But we also would be foolish to assume that we—or our world—can never be vulnerable, and that we will always win every time we do battle with the odds. Delusions of immortality may die hard, but they all do die.

Whether the probabilities are overwhelmingly in our favor or much less cheerful, it is practical and responsible to weigh the risks and realities with as much precision as possible. When we preemptively relegate a danger to the trash can labeled "unthinkable," we literally refuse even to think about it...but that doesn't mean it can never happen to us. We might well prefer not to dwell on our own death, and consider our own demise unthinkable, but if we take that attitude as an excuse to avoid buying life insurance or making a valid will, we do a terrible disservice to everyone who depends on us.

I'm not suggesting that we make the worst-case scenario our benchmark for all environmental decision-making. I sometimes joke that all of my books on environmental threats could be boiled down to a core essence that fits neatly on a bumper sticker: BETTER SAFE THAN SORRY. Like most bumper stickers, though, its fundamental truth comes bundled with a generous helping of oversimplification. Absolute, complete safety is an impossibility, utterly unattainable no matter how elaborately and extravagantly we construct our bubble of security. And as we approach the maximum degree of safety, we also find the costs of achieving it rocketing toward infinity, both from lost opportunity costs and the actual price tag of the preventive and remedial measures we implement. The quest for total safety in a worst-case-scenario context ends with the highest possible cost for the lowest possible activity. It's the polar opposite of the mantra for the quality movement; rather than doing "more with less" we wind up doing less with more. In the world of the worst-case scenario it is forever winter but never Christmas.

An irrationally risk-averse, safety-addicted mindset produces the paradoxical, and counter-intuitive, effect of making its adherents at once more vulnerable (or at least vulnerable in a different way) and less productive. Inactivity and excessive caution have their own costs, and these too should be considered and weighed against the benefits and detriments that come with the alternatives. This evaluation, within any realistic framework, necessarily involves an informed, objective, rational, and logical examination of the relative risks and rewards associated with *each* option. This balancing act must include full assessment of risks and rewards accompanying the option to take *no action* other than to maintain the status quo. It also has to incorporate our best estimate of the probabilities of any particular outcome actually happening, for every possible option. But sadly, when it comes to the main topics of this book—global environmental threats—the qualities of realism, objectivity, rationality, and logic, as focused on a sound foundation of evidence and information, are in short supply. On the contrary, the predominant mode of debate within this most important sphere tends to be of the type exemplified by the following statement: "Shut up!" he explained.

There often is no real debate on monumental issues such as climate change and mass extinction. Instead, people take sides and line up in opposition to one another, like warring armies, and denounce one another. Like George Orwell's "two minutes hate," the opposing camps' arguments in our real-life brave new world are much stronger on vitriol and volume than on reason and rigor. When people are unwilling or unable even to acknowledge the existence of any evidence supporting a contrary viewpoint, or the good faith and honorable intentions of people with a differing opinion, the arguments easily descend into something resembling a holy war.

Where evidence, perspective, relative probabilities, and analysis of alternatives are supplanted by a quasi-religious insistence on a particular brand of dogma, conditions are ideal for the type of reasoned, collaborative decision-making typified by the crusaders and the jihadists. Everyone feels that they have a dogma in the fight. Nuance gives way to false certainty, and the self-styled enemies are much quicker to reject than to respect the ideas from "the other side." As with any type of holy war, environmental crusaders and jihadists find it very difficult to change anyone's mind except by good old-fashioned forcible compulsion. Facts and figures only get in the way when both sides "just know" they are right and the others are stupid and wrong, if not downright evil.

It is irrational for us to refuse to consider, as objectively and thoroughly as possible, the merits of opposing opinions, relative probabilities, comparative risks/benefits, and underlying factual foundations in the biggest environmental issues of all time. We might look at the situation objectively and shake our heads at the utterly self-destructive stupidity of it all. But it's far from the only example where people behave irrationally in making important decisions. In fact, it is very similar to the way many people approach some common daily activities and familiar games of chance where they have a personal and very real stake in the outcome. Let's consider how we are literally betting the Earth in the same way (for better or, you guessed it, for worse) we bet our money and sometimes our lives in our individual probability-dependent activities. For the remainder of this chapter I will analyze some everyday-life examples that don't explicitly involve what we commonly think of as gambling. Then, in the following chapter, I'll turn to the more obvious and traditional forms of gambling and the role of probabilities and uncertainty in making more or less rational decisions within that environment.

If you didn't get hopelessly lost during our whirlwind tour through the foggy paths blazed by Blaise Pascal, Werner Heisenberg, and Kurt Gödel—or even if you did, for that matter—you should now have some grasp of the extent to which probabilities affect all of reality. If your image of this constant influence of probability more closely resembles a Monet than a Michelangelo—more blur and blend than sharp, razor-clean boundaries—then that is just what I'd hoped to achieve. My law students are all too familiar with their eccentric professor's relentless insistence that the first and best answer to any legal question always begins with the timeless axiom, "It depends." But it's true. It *does* depend, always, on a variety of variables…in the law, and in all of life.

Our lives, and the vastly wider universe in which we reside, are in a perpetual freestyle dance with the unseen partner named probability. We might prefer to think of ourselves as masters of our own fate, hands firmly on a steering wheel that responds perfectly to our slightest nudge in one direction or another. But this notion is a myth, the twin sister of the outmoded view of the atom as a central solid ball orbited by smaller balls moving in an obedient and orderly, well-defined route around the core. We now know, from the insights of Heisenberg, Bohr, Einstein and others, that an atom is a much fuzzier glob of imprecision, with the relative positions of nucleus and electrons at any given moment smeared and smudged into a mushy, cloudy fog of a probability distribution. Viewed from afar, the atom would look more like the planet Venus, forever snuggled within a smoggy electron cocoon, than sunny, clear Mars.

So, too, on a much larger, more recognizably life-size scale, the forecast always reads "Partly cloudy with occasional patches of dense fog." We might indeed have our hands on a steering wheel, and we might take great care to turn the wheel in exactly the direction we want at precisely the instant we think best, but there is a design defect built into the steering mechanism. Our vehicle, an antique model *Reality*, will respond to our attempts to steer with a somewhat predictable, somewhat uncertain range of reactions. Maddeningly, and to our frequent dismay, our exertions to steer end up losing something in the translation. They're interpreted or misinterpreted by our car as rather vague suggestions rather than sharp, crisp military commands. Stuff happens, and it tends to happen with a liberal sprinkling of random chance added to our careful planning. We do have significant influence over the direction of our lives, but as Mark Twain once quipped, nothing is certain but death and taxes. And even death is uncertain as to the exact time, place, and circumstances, while taxes are also subject to the whims of our political leaders, our income, the tax commissioner's aggressiveness, our accountant's ingenuity, just plain luck, and the extent to which we are willing to take a chance on a change of residence to the nearest federal penitentiary.

Robert Burns knew this long before Pascal, Gödel, and Heisenberg collectively rained on our meticulously organized parade. The Scottish poet wrote in 1785 about how the best laid schemes of mice and men "gang aft agley." We may not say "agley" very much in our casual conversation these days, but we still know instinctively that (a) it isn't a good thing; (b) Burns was right; and (c) we've frequently seen our own plans careen off the road into that ugly agley ditch.

Mark Twain and Bobby Burns were commenting on an aspect of the human condition, and of everything else we see, that people have understood on some level for thousands of years before Gödel thought great thoughts about unprovable truths, or Pascal blazed the trail toward probability theory, or Heisenberg conceived of a universe of universal uncertainty. In a very real and permanent sense, everything is a gamble, and everything has an element of risk, potential gain or loss, and odds. Sometimes this gamble is obvious and up front, in our face and plain to see, and sometimes it is well disguised under several layers of camouflage and costume. But it's always there. The answer always begins with "It depends."

Let's briefly consider several examples of gambling (or, if you prefer, probability-based activity) most of us can relate to on a personal level. We will begin with some personal activities that involve decision-making in an environment of uncertain and incomplete information. For some readers, this may be the first time they've considered these choices as gambles at all. Many of you might have never thought about the ways in which these personal options inherently involve risk, reward, uncertainty, and a weighing of them all in light of probabilities. Still, I will try to show that, perhaps on some subconscious level, we do deal with this entire list of factors every time we encounter any of these familiar situations (and many others like them) in our lives. Then, in Chapter 9, we'll move on to an analysis of more standard and overtly probabilistic, money-on-the-line forms of gambling.

In every instance, I intend to show that we are in fact engaging in the type of rational decision-making I've presented in this book, even if we are not consciously aware of it, and even if we are not really tackling the issue logically. I will try to demonstrate how my World Wager in some form could apply to each situation, and how failure to apply it can lead to very unfortunate consequences, even for a single individual. This discussion will then set the stage for a look at how all of these same principles are also at work on a global scale when we take on the world's biggest environmental questions...only at much, much higher stakes. So fasten your seat belts once again, because it's going to be another rough ride...and we wouldn't want to take any unnecessary chances, would we?

a. Driving.

Unless your driving skills are as bad as mine, you don't think of sliding behind the wheel of a car as a gamble. But then, before you began to read this book you didn't think of ordinary human activity (or inactivity) as the equivalent of betting the Earth either. Let's break down some of the relevant characteristics of driving and examine the ways in which operating a motor vehicle might in fact be a microcosm of the global issues challenging the world today. By this thought experiment I don't mean to encourage you to cultivate grandiose delusions (although I admit that they are enjoyable, speaking from personal experience), but merely to illustrate the complex interplay of known and unknown variables and elusive probabilities underlying even the most commonplace of human activities.

If we reflect on what is typically involved in driving, we can list the following *facts*, among others, as those *generally accepted as true*.

- Most driving experiences do not result in any accident.
- Some driving experiences result in property or physical damage to one's own car and passengers.
- Some driving experiences cause property or physical damage to another person's vehicle, other property, and passengers.
- People are sometimes hurt or killed when driving a car.
- People are sometimes hurt or killed by someone else driving a car.
- Some types of vehicles are generally more prone to suffer serious damage in the event of a crash, or more likely to be involved in an accident, than others.
- There are factors that influence both the likelihood of vehicular accidents and their severity.
- These factors include road conditions, weather, lighting, amount of traffic, vehicle safety (for all vehicles involved), and driver characteristics (for all drivers involved).
- A driver's physical health, degree of alertness, level of fatigue, amount of driving experience, ability to concentrate on driving, and impairment from alcohol, medications, or drugs can significantly affect his or her capability to operate a vehicle safely on any given occasion.
- Some serious defects in our vehicle and in other vehicles on the road are difficult or impossible for most people to detect before it's too late to avoid disaster.
- Safety features such as seat belts, if used, can help reduce the risk of serious injury or death in some percentage of accidents.
- There are some types of accidents in which the use of seat belts can actually increase the risk of serious injury or death.
- Air bags can also increase safety in some situations, while worsening risk in others.

- Irrespective of other factors, the way a given driver operates his or her vehicle on a particular occasion (speed, aggressiveness, anger, carelessness, suddenness of turns, etc.) also influences the likelihood of an accident.

This is, of course, a very incomplete list of facts associated with the risks of driving, but it will be good enough as a starting point for purposes of illustration. We can supplement it now with a litany of variables relevant to the items in the above list, each of which will (we can all agree) make those facts somewhat more or less likely to add up to a safe driving experience. Of course, these variables can also combine together, in any number, to *influence* the outcome of any decision to drive on any given occasion.

- Road conditions.
- Weather and other visibility factors.
- Size, structural design, and weight of the vehicle(s) involved.
- Fatigue (of ourselves and other drivers on the same road).
- Sobriety (of ourselves and other drivers on the same road).
- Driving skill (of ourselves and other drivers on the same road).
- Distractions (of ourselves and other drivers on the same road).
- Physical health (of ourselves and other drivers on the same roads).
- Alertness (of ourselves and other drivers on the same roads).
- Time of day.
- Existence of special risk factors (e.g., holidays where excessive drinking is widespread).
- Visual acuity (of ourselves and other drivers on the same road).
- Emotional state (of ourselves and other drivers on the same roads).
- Traffic congestion.

- Driving speed (of our own car and other vehicles on the same roads).
- Duration of our drive.
- Lighting conditions.
- Operational defects in our vehicle (and in other vehicles on the road).
- Familiarity with the route (for ourselves and other drivers on the same roads).
- Presence, type, size, and location of hazards or obstacles in the road.

We tend to be more or less aware of the above variables as they apply to ourselves or our own vehicle; but we *can't* know the situation with regard to any of the dozens, hundreds, or thousands of other drivers with whom we will share the road during any particular drive. The conditions and tendencies of those other cars and drivers are very important and relevant to an accurate calculation of the probabilities that our trip will be safe and uneventful, but the best we can do is come up with a reasonable guess on any of them. Most of the time, of course, we perform this type of calculation subconsciously at best, and don't give any real thought to whether it is a wise decision for us to take the wheel. We subliminally assume that all the risk factors are minimal or nonexistent and there is no reason to hesitate or weigh the alternatives before we drive our car. This involves a personal, conveniently travel-sized variation of my World Wager, in which even *devastatingly severe possible outcomes* (such as the deaths or disabling injuries of ourselves, our passengers, and other people) are deemed not so *probable* as to curtail all driving activity, but only to generate a normal or moderate amount of care, precaution, and caution. Within any "normal" risk-acceptance framework, this is in fact a reasonable and rational outcome under ordinary circumstances. This is not to say that there is any absolute guarantee that we will drive safely on any given day, but only that we are prudent in concluding that ordinary care is all the situation calls for.

However, under special circumstances where we know that we or our own car are not up to par, our risk/reward calculus can (and should) become much more conscious and deliberate, and can often result in a different rational decision. Examples of such heightened risk factors include any situation where we have reason to know that we are significantly impaired in our ability to drive, whether due to consumption of alcohol, medications, or drugs, illness, emotional upset, tiredness, loss of eyeglasses, physical pain, or excessive distractions. Either alone or in combination, these variables can dramatically alter (and increase) the probabilities of negative outcomes, and thus flip the rational decision to one in which we either stay put or get someone else to do the driving for us. The same is true for factors affecting the safety and operability of our vehicle itself, such as bad tires, faulty brakes, poor steering mechanism, etc.

Note that in any set of circumstances, we can never know for certain what the actual probabilities are or what the severity of any unfortunate result would be. We may truly have almost no idea of the real likelihood of serious injury or death to ourselves or others during any specific drive, be it one chance in 10,000 or one chance in 10. We'd be at a loss to guess what the true numerical odds might be, or even to come within the correct answer by a factor of 10, but then we don't really have to be able to put an actual number on the situation. All we can ever have available to us is a fairly wide range of assumptions about outcomes and odds, with a very healthy allowance for leeway and variability—a ballpark estimate of whether it's reasonably safe on any particular occasion for us to drive, all things considered. It's a big park at that, more like Yellowstone than Wrigley Field. But it's one within which we intuitively feel that we know enough about all the most important factors to make a reasonable decision on the simple but potentially life-and-death question: Is it safe enough for me to make this drive right now?

I need to add one other point, and it's something that will come up again and again in the last chapters of this book, because it constantly recurs in every choice we make. That point takes aim at the pros and cons, risks and rewards, and probabilities associated with the *alternatives* to taking action. In this example, if we are trying to decide whether to drive (either now or ever), our analysis would be terribly incomplete and inadequate unless we also take into account what it means to us if we *don't* drive. What good and bad results might flow from a decision *not* to make this trip by car? What happens if we simply stay where we are? Would we miss some crucial opportunity? How safe, affordable, convenient, adequate, economically productive, and effective are the other forms of transportation available to us? Inevitably, there are some risks that come along with every decision not to act, just as with every decision to act.

I'm going to discuss this principle later, but I want to introduce it now because it is so easy to overlook. In Chapter 10, I will explain it in considerable detail within the context of what I call Type P and Type R errors; for now, let's limit our discussion to the basic idea.

Here's the foundational notion. It often *seems* safe to do nothing, to refrain from taking any given action, to stand pat...because when we take that "no action alternative" we don't appear to be changing the status quo in any way. So where's the risk, Chicken Little? The fact is that there is indeed a danger in taking the "no action alternative," even though it may be cleverly disguised as comfortable, familiar, nostalgic home-sweet-home. Every time we choose not to act (whether by "act" we mean drive, take medicine, exercise, hang-glide, or do anything else that requires a change in our current behavior) we avoid the risks that could come with action, but we also necessarily accept the risks that come from inaction. We accept the danger that we will miss the benefits that might have come from action, and we embrace the hazards associated with the alternatives to taking action. Yes, they also gamble who only sit and wait. For now, just keep in mind that there's no such thing as a free lunch, or a risk-free decision, even a decision to do nothing. Period.

Blaise Pascal himself recognized this point (among many others) in his original Wager. He was characteristically pithy in stating this crucial insight in just a few powerful words. Pascal anticipated the objection from some skeptics that they would prefer not to bet one way or the other on God's existence, thank you very much, and would rather just sit this one out. Here is Pascal's rebuttal: "But you must wager. There is no choice. You are already embarked." He recognized, and even took as an obvious truth, the fact that simply by being alive we are in fact wagering every minute, either by believing in God or not. The same holds true in general, for any decision that asks either for action or inaction, active belief or its absence, motion or stasis. Every time we sit out a dance, we're deciding not to dance. It's amazing how many choices we can make just by doing nothing. Personally, I'm making whole bushels full of decisions right now, even though it might appear that I'm merely loafing. Human beings are veritable marvels of prolific apathy.

b. Smoking.

We might perceive a more obvious link to gambling in a decision whether to begin smoking cigarettes than in pondering whether to drive. This could be attributable to the greater attention devoted to the dangers of smoking than ordinary driving, but in fact the process of arriving at a rational choice is very similar. Now, I am not claiming that anyone actually goes through this type of analytical process when considering whether to take up smoking. I've never heard of anyone who brainstormed all the possible pros and cons, analyzed the odds, considered alternative activities, made a decision matrix, and methodically calculated whether adding smoking to her daily to-do list was a positive Expected Value choice. It's one of those things in life that just seems to happen, or not, with little or no thought about benefits and detriments and probabilities. All I am trying to do is show how a reasonable person *might* incorporate uncertainties and probabilities in arriving at a sound decision for something that may be a matter of life and death.

Let's again list just a few *relevant facts* about smoking, to which most intelligent and reasonably well-informed observers would agree. Here are some givens that pertain to the question, "Should I smoke tobacco products?"

- There is some evidence that smoking increases the risk of cancer.
- There is some evidence that smoking increases the risk of emphysema and other respiratory disorders.
- There is some evidence that smoking increases the risk of heart disease.
- Each of these medical conditions can also be caused by factors other than smoking, including innate genetic traits and lifestyle habits.
- It costs a significant amount of money to buy and use tobacco products and related materials.
- Many smokers report that they enjoy smoking, whether because of social interactions with other smokers, or a pleasant physical sensation, or the comfort of regular ritualized behavior, or other apparent benefits.
- Some smokers find that smoking helps them to relax, or to concentrate on important tasks, or to keep from gaining weight.
- The extent to which smoking causes health problems depends to some extent on genetics, age, exposure to other influences, general health, etc.
- People who smoke heavily tend generally to be more at risk for health problems than those who smoke less.

- Some tobacco products present a greater health danger than others, due to factors such as filters, amount of nicotine, quantity of tobacco smoked, direct contact with lips, etc.
- There are other effects associated with smoking that are not explicitly health-related but may have social and quality of life consequences, such as reduced stamina, discolored teeth, smoker's breath, nicotine stains, less discretionary money available to spend on dates and other social events, and psychological trauma from being regularly exiled to the outdoors in order to smoke.

There are also *unknowns* tangled up in the smoking equation that will exert an effect on our decision-making process. No one can know with any degree of certainty the value any of these actually takes in any particular person's case, but we can intuitively appreciate their relevance and their elusiveness. They include the following deck of wildcards, among others.

- Our specific genetic predisposition to contracting cancer, emphysema, or heart disease.
- The extent to which this predisposition will interact with varying degrees of exposure to tobacco smoke.
- The presence of other serious latent health problems that may pose a greater and more immediate threat than tobacco-related problems.
- The probability that our intake of second-hand smoke will cause health problems independent of whether we ourselves smoke.
- The results that flow from a combination of smoking and our other life choices, such as drinking, overeating, lack of exercise, poor diet, inadequate sleep, use of legal or illegal drugs, conflict-ridden personal relationships, and a stressful occupation.
- How much we will smoke.
- How long we will smoke.

- Whether we will be able to stop smoking if we want to.
- Our risk of dying from other causes such as occupational accident or latent genetic defect.
- Our individual aversion to the particular health problems usually associated with smoking.
- Financial costs in the long run of buying cigarettes in the quantities desired, compared to other competing demands on those resources.
- Whether we will gain a significant and physically harmful amount of weight if we don't smoke or stop smoking.
- Whether we will suffer other serious negative effects if we stop smoking, such as increased nervousness, inability to sleep, lack of concentration, heightened irritability, etc.
- Adequacy, efficacy, side-effects, and affordability of the medical treatment/cure for the various smoking-related maladies.

A person hypothetically contemplating the addition of smoking to her list of recreational activities could, on some level, think about these facts and factors, together with at least a rough probabilistic weighting of the relative pros and cons of each course of action. For a person who is already a smoker and is trying to decide whether to stop, they would also need to add variables relating to the cost and efficacy of available stop-smoking aids, the benefits of quitting after already smoking for a certain number of years, and any possible adverse effects from quitting (such as weight gain and loss of emotional comfort).

Using a variation of the World Wager and decision matrix, a rational person would need to weigh the worst possible personal harm from a decision to smoke (i.e., a lingering, agonizing death) against some estimate of how likely it is that this harm would befall them and what it might cost to avert that harm. There are intangibles involved, such as the degree of emotional, physical, or psychological pleasure people derive from smoking. There are huge and unpluggable gaps in the facts, such as the probability that any given person will contract a deadly disease at some point in time if they smoke to a certain extent for a particular duration. There are major unknowable unknowns relating to our individualized genetic predisposition to smoking-related diseases and other life-threatening causes. There are also indelible question marks surrounding the extent to which any given person's exposure to other stressors may be more significant than smoking as a risk factor.

This involves a lot of guesswork, at least on the subconscious level. It is, in any event, a form of gambling, and high stakes gambling at that. We don't really know the odds we're facing, so we are forced to generalize and extrapolate from incomplete information and data from other people's situations. How much does a given amount of smoking increase our preexisting chances of getting one of these dread diseases? No one knows for sure. But we do have some idea of how risk-accepting we are, and how important it is to us to reduce (but not eliminate) our chances of dying from certain diseases. Again, there are no guarantees. Even if we never smoke, we could nonetheless fall victim to cancer or emphysema or heart disease, none of which exclusively targets smokers. Likewise, even if we smoke heavily for many years, there is no certainty that we will contract any of the smoker's maladies. It's a matter of odds and oddities, probabilities and proclivities.

What does the "no action alternative" look like in the smoking context? If we are already a smoker, the status quo decision would be to continue smoking the same product in the same quantity. The risks from taking no action to change that behavior are the ones typically associated with smoking, while the benefits are primarily continuation of whatever positive feelings and enjoyment we derive from tobacco use. But if we are currently a non-smoker, the no-action choice is to remain that way and not begin smoking at all. Here, the status quo decision avoids all the added risks that come from being an active smoker, although it can't guarantee that we'll never be stricken with cancer, heart disease, etc., from some other origin. The risks from not taking up smoking may pale in comparison, but they might encompass weight gain, heightened nervousness, emotional restlessness, and resort to other potentially harmful compulsions as a substitute for whatever impulse is urging us to begin smoking.

This serves to illustrate that, although there is a gamble built into every decision to act or not act, this in no way implies that both alternatives are equally risky! There is some risk from not smoking, just as there is an element of risk from becoming a smoker, but the evidence tends to show that it is far more dangerous to smoke than not to smoke, all else being equal. The choice is in the form of an either/or proposition (smoke, or don't smoke), but the two options are of very dissimilar risk in terms of size, nature, and probability of the consequences. This may seem entirely obvious to you, but people do make the mistake of looking at such a binary, yes-or-no situation (either we win or we lose) and concluding that, because there are only two possibilities, the odds must be 50/50! I love to play poker against these mathematically challenged individuals.

One need not have ever heard of Gödel, Pascal, or Heisenberg to grasp that the smoking decision is a matter where unknowns lurk everywhere we look, and much hinges on how we view that doubt in light of the upside and downside potential inherent in any course of action. If we view our personalized, individualized, incremental increase in risk of deadly disease as sufficiently small, in conjunction with the money it costs to buy cigarettes and compensate for any moderate health effects, compared to the loss of enjoyment from smoking and possible weight gain and increase in irritability, we may decide to begin or continue smoking. This is likely to be calculated mostly or even completely on a subconscious level, with occasional bouts of conscious consideration under the stress of a bit of news or new information, but it still happens. It is a game of chance, whether or not we know it. Objectively, the decision on smoking has a significant chance of affecting the duration and quality of our life in a major way. If it surprises you that we make such personal life-threatening decisions with this much carelessness and thoughtlessness, stick around to see what we're doing in matters that could endanger the whole world.

c. Flying in a commercial airliner.

When we drive our car, we are taking a gamble for which we have significant, though by no means exclusive, influence over the outcome. The situation is in some ways similar, but in major respects very different, when we are faced instead with the decision whether to take a flight on a commercial airliner. You are beginning to see the point, I'm sure. There is an element of gambling and uncertainty in *every* human activity and in *every* choice, whether to act or refrain from acting. Let's think about the usually-hidden issues underlying any situation where flying is an option.

As is generally the case, there are some salient facts about flying that any reasonable person would agree are true. Here is a partial list.

- Most commercial airline flights do not end with anyone being hurt or killed.
- Some commercial airline flights do crash, killing everyone on board.
- Some airplanes are safer than others, all else being equal, because of size, newness, quality of design and manufacture, number and type of engines, etc.
- Some airlines are better managed than others in terms of factors that affect safety.
- Some professional pilots are more highly skilled and safer than others.
- If you need to travel, alternatives to flying also involve some risk of serious injury or death.

- Wind, precipitation, nearby birds, poor visibility, ice, lightning, and other environmental factors can affect the safety of a given flight.
- The safety of a flight also depends on proximity to other aircraft, runway conditions, obstacles in the flight path, quality of guidance from air traffic control, etc.
- The extent to which an airplane has been recently and thoroughly inspected and given high-quality preventive maintenance influences its safety.
- Air travel is often the fastest means of transportation between two points, depending on the distance and barriers between them, convenience of airports, availability of alternate means of travel, and other factors.
- The pilot and co-pilot will be less capable of safely operating the airplane if they are fatigued, ill, under the influence of alcohol or drugs, emotionally upset, etc.
- There is some chance than any airplane will become involved in a terrorist attack along the lines of those of September 11, 2001.

Following our now-familiar pattern, some unknowns also exist that are both relevant to any decision of whether to fly and elusive of precise resolution regardless of the amount of research and effort we expend on finding the answer. Here are some of the most important such variables in our "To fly or not to fly" equation.

- The exact set of weather and environmental conditions that will confront any particular flight.
- The probability that existing weather and environmental conditions at any phase of our flight will cause a crash.
- The exact set of weather and environmental conditions that will confront any alternate means of travel to the same destination.
- The probability that terrorists will target our specific flight, and will succeed in doing so.

- The extent to which alternate means of travel to the same destination are more or less safe, affordable, timely, feasible, and convenient.
- The probability that any significant hidden defect in the airplane will cause a crash on any given flight.
- The probability that the pilot or co-pilot will be able to handle any unforeseen threat to the flight.
- The precise cluster of conditions currently affecting our pilot and/or co-pilot.
- The likelihood that another aircraft will collide with ours, whether in the air or on the ground.

Once we begin to analyze the situation in this manner, it quickly becomes apparent that it is a very complicated decision fraught with multiple important variables for which we can never know the precise value. This may be why most of us never think this way about flying. We sense that if we did, we would never go anywhere near an airport...except then we also would need to consider the implications of other means of travel, or staying home for that matter. There is danger and uncertainty in each of these alternatives, and in every alternative to any possible course of action.

What do we know for sure about our decision on flying? We know that no alternative is completely guaranteed to keep us 100 percent safe; even if we stayed home with the door locked and the blinds closed, we could slip on a wet bathroom floor and suffer a fatal injury to our head! We know that the likelihood of most of the devastating potential occurrences on our list, while more than zero, is (thankfully) extremely small. We know that for many trips, there really is no adequate and acceptable substitute for flying, because of the distances involved, the need to arrive quickly, the difficulty and danger of doing a large amount of driving under time pressure, the requirement to cross oceans, and other practical considerations. In other words, although there are always theoretical alternatives to flying, we aren't always prepared to accept them due to what it would "cost" us in terms of time, inconvenience, quality of life, added difficulty, lost opportunities, other dangers, etc.

We cannot completely rule out a fatal incident, no matter what choices we make or how much we study all the variables. We could select the finest, most well-maintained aircraft from the airline with the world's best safety record, operating with the most rigorous set of anti-terrorist security mechanisms, and wait to fly until all data indicate the weather will be perfect; but still there would be some non-zero probability that all of our efforts would end in a flaming heap of shattered rubble on the ground. We could also decide that flying is just too dangerous, and either drive, or take a train or bus or ship, or simply stay home…yet again, there are no air-tight guarantees that we will survive the experience. Stuff happens, and it happens no matter who we are and what we decide to do. Comforting, isn't it? Welcome to *my* world, and thank you for betting on it.

d. Immunizing our children.

Even actions that are supposed to make us safer and healthier are not exempt from the Grim Reaper, disguised as Lady Luck. We might opt for some lifestyle choices (such as a program of regular, vigorous exercise) with the intention of thereby prolonging and improving our lives, yet end up causing diminished health or even premature death instead. The same is true for the medical choices we make for ourselves or our children. No drug, no inoculation, and no surgical procedure is guaranteed to be 100 percent effective and universally safe in all cases, under all sets of circumstances, and for all people. That's why we are bombarded with those endless, breathlessly rushed warnings and disclaimers that take up the last 75 percent of every advertisement for every pharmaceutical product. And I personally figure that they wouldn't take the time to warn us about all those horrible possibilities if there weren't a good reason, so I never, ever buy any of those medicines...which I'm confident is not the advertisements' desired effect.

An important example deals with any parent's decision whether to have his or her child immunized. Beginning in our children's infancy, we are confronted with a lengthy array of inoculations available to immunize them from such potentially devastating, even fatal maladies as chickenpox, rubella, polio, measles, mumps, diphtheria, tetanus, pertussis, and hepatitis B. The government goes to considerable lengths to persuade parents to have their children inoculated as a preventive measure against all of these medical threats. The idea is that kids who receive these inoculations initially, and any recommended follow-up booster shots, will be much less likely to be stricken with the associated medical problems, and to pass their illness on to others. Further, if they do become so afflicted the assumption is that they will only have to endure a far milder case than they might otherwise suffer.

You know the steps to this dance by now. We'll begin by listing some facts that would seem to be beyond rational dispute on the question of childhood immunizations.

- ♦ The diseases and other medical conditions known as polio, rubella, tetanus, hepatitis B, chickenpox, measles, etc., can pose a serious health risk to either the child who contracts them or to those with whom the child comes into contact, or both.
- ♦ A significant number of children (and even adults) have died, or suffered serious, life-changing damage, from each of these medical threats in the absence of immunization.
- ♦ The inoculations are widely available either at no or very low financial cost to the individual.
- ♦ The inoculations intended to protect against these medical threats are effective and safe in a very high percentage of cases.

- Failure to immunize can sometimes cause administrative obstacles to arise in dealing with the government, including the public school system.

- The inoculations intended to protect against these medical threats cause no significant harmful side effect to the recipient in a high percentage of cases.

- There is some evidence (of disputed credibility) that some very serious, life-changing side effects, such as autism, may result from the inoculations in some small percentage of cases.

- Because of the prevalence of inoculations for many years in many parts of the world, the probability of an un-immunized person encountering someone with one of these diseases in an infectious stage is low, but greater than zero.

I want to highlight the fact that I included a reference to evidence of disputed credibility among the list of givens. Frequently, we can find evidence on more than one side of some essential elements of a decision we need to make. I'll return to this point in Chapter 10, but think about how pivotal it will be to our decisional process whether we consider evidence like the autism-inoculation link to be a proper part of the inquiry. The resolution of that threshold evidentiary matter can directly alter the overall result of our analysis. As lawyers like me tend to say, this issue can be outcome determinative.

The butterfly effect is fluttering all around us as we think about whether to factor in arguably questionable research findings. Big differences in outcome can easily follow from an initial decision to credit some controversial findings. Our threshold choice can serve as a filter or a gatekeeper, controlling how the evidence stacks up for or against the proposition at hand. Depending on the way we sort it out, we can end up with evidence that seems to overwhelmingly favor one result, or evidence that appears to be roughly equal on both sides. We can think of the problem as one of admissibility of evidence, or at least a matter of how much differential weight to assign to various pieces of evidence. The answer to the battle of what evidence is worthy of consideration, and whether some admissible evidence is "more equal than others," will often give us the answer to the overarching question before us. So keep this in mind until we revisit it in the final chapter.

What are the unknown quantities applicable to any parent's decision to immunize, or not to immunize, his or her child? These are among the list of question marks firmly and stubbornly attached like permanent barnacles to our decision-making ship.

- ♦ The precise probability that any specific child would ever contract polio, rubella, tetanus, hepatitis B, chickenpox, measles, etc., if he or she is not given the associated inoculations.
- ♦ The degree of severity with which any specific child would be afflicted with these medical conditions, if he or she contracts them at all, in the absence of the associated inoculations.
- ♦ The probability that our child, if left without immunization, will contract and spread any of these medical conditions to other unprotected individuals, and the level of harm this would inflict.
- ♦ The exact probability that any specific child who receives any or all of the recommended inoculations would suffer a significant side-effect.
- ♦ The probability that any such significant side-effect would be of major, life-changing dimensions, such as autism, for any specific child.
- ♦ The relative costs in money, diminished quality of life, reduced capacity, lost opportunities, shortened lifespan, etc., from the likely consequences of actually contracting one of the listed medical conditions as opposed to the most probable adverse side effects.
- ♦ The probable broader impacts on other people/society from a decision to immunize or not, in the individual case.

♦ The probable broader impacts on other people/society from a decision to immunize or not, in the case where many people independently arrive at the same decision and the individual effects are multiplied manifold.

This example is one that affects a lot of people in actual practice. Intelligent, reasonable, well-educated people regularly make decisions on both sides of the immunization debate, every day. This tells me that for a fair number of people there is at least the *perception* if not the actuality of significant probabilities that: (a) the listed medical threats (measles, polio, etc.) are not sufficiently serious and/or not sufficiently likely to be contracted to warrant the risks from the inoculations; (b) the possible severe and irreversible side-effects from inoculation (e.g., autism) are high enough to dictate caution; and (c) the combination of (a) and (b) in light of their comparative likelihood and magnitude points in the direction of not immunizing. Even if the reality is very different from the perception, perception alone can be enough to move people toward momentous personal family decisions—in this case, to go with the "no action alternative" and refuse to immunize their children. And this happens even though there are well known and potentially severe risks linked to the no-action option.

This immunization example has some significant parallels with global environmental decisions regarding climate change and biodiversity loss. In each of these situations, we need to decide whether to implement a preventative course of action (that itself comes with its own considerable costs and risks) in dealing with a threat that is potentially extremely serious but also of uncertain real impact. In each instance, there is the question of whether the cure may be worse than the disease, or, to put a new spin on yet another cliché, whether a pound of prevention is too much to devote to something that may only need an ounce of cure, but *might* also need a ton, and might even be incurable. This reflects the fact that there is in each case the danger (with some unknown probability) of suffering catastrophic harm if we *fail* to take the precautions at hand and end up having the worst-case scenario smash us in the face. But we're also playing a high-stakes game of chicken against the opposing risk of disaster if we *do* take precautions and, ironically, they explode in our hands.

A worried parent searching for guarantees on the immunization issue is going to find them as scarce as a Dodo's teeth. No inoculation is 100 percent safe. The odds may be stacked heavily against any given child suffering a catastrophic side effect from the immunization, but there is nonetheless at least some controversial evidence that there's a real, non-zero probability that such a tragedy could happen. Do we dismiss the disputed claims entirely and, if not, how much weight are they worth on our conceptual balance? Plus, no inoculation is 100 percent effective, so even an immunized child could possibly still end up suffering from the feared conditions (measles, polio, and the like).

And sadly, the threats to life and health that these inoculations are designed to reduce are still out there, and still as potentially devastating as ever. They may well not be nearly as prevalent as they once were, thanks in large part to the efficacy of widespread immunization, but they have not been wiped out, not even from the so-called developed parts of the world. There is no guarantee that a child who has not been immunized will be fortunate enough to steer clear of all of these medical hazards for an entire lifetime. And to the extent that progressively greater numbers of parents opt out of the inoculations, it is likely that these hazards will grow increasingly more common and more dangerous. If skepticism moves enough people into the no-thanks camp, polio and the other perils could mount a major resurgence, and this would further influence the variables that drive our decision.

Under these circumstances, a rational decision-maker will need to make some preliminary but difficult choices as to his or her acceptance of or aversion to risk, the relative probabilities of each possible outcome, and the comparative magnitude of harm that lurks at the end of each available dark alley. Only after making those threshold determinations can anyone be prepared to make a reasoned and logical decision on the appropriate choice for his or her own individual situation. Many people will go through this process without any awareness that this is exactly what they are doing; but they will still engage in at least some variation of this type of analysis on some subconscious level.

How else could we make this kind of difficult, maybe-damned-if-you-do and maybe-damned-if-you-don't decision? We might just trust it all to random chance, relying on a Magic 8-Ball or a few rounds of eenie-meenie-minee-moe. We could consult an astrologer, tarot card reader, or soothsayer to soothe our concerns. We might pray about it and try to be receptive for spiritual promptings, or, on a less elevated plane, just "go with our gut." But to the extent we invest any of our own thought and intuition into the inquiry, it seems inevitable that we are on some level weighing advantages and adversities with some adjustment for relative size and likelihood of each. How about one more tinkered-with cliché? What's good for the goose is good for the gander, and what's good for one person is good for the planet.

e. Exercise.

The question of whether to exercise, physically, is another example of "gambling" we all must encounter in our everyday lives. Similar to the issue of whether to immunize our kids, the decision about exercising also has strong links to the far broader, planet-wide questions that are the subject of this book. These matters involve the option of doing something proactively that might avert a potentially ruinous tragedy, but with costs and benefits, probabilities and uncertainties, all around. I was tempted to call this sub-chapter *Betting the Girth*, but then my natural good sense took over and I beat back the temptation.

What are the basic foundational facts we might all accept as givens relevant to the choice we face on exercising? Here is a list of some pertinent and fairly indisputable truths.

- Some level of physical activity or exercise can usually help to maintain or improve a person's health and well-being.
- Various forms of exercise primarily affect flexibility, strength, cardiovascular fitness, and endurance.
- Some forms of exercise tend to have a greater beneficial impact than others.
- Some forms of exercise are more strenuous than others.
- Some forms of exercise tend to have a greater risk of serious injury or death than others.
- There are factors that affect a given person's need for and ability to engage in various forms of physical exercise.
- Some people have hidden medical conditions that might make it very dangerous for them to engage in certain forms of exercise.
- Lack of exercise over a prolonged period can lead to dangerous levels of obesity, loss of muscle strength, loss of bone mass, reduced cardiovascular fitness, lower energy level, reduced stamina, diminished ability to concentrate, and increased risk of heart attack or other serious health problems.
- The degree of difficulty, type, duration, and regularity of an exercise program can affect both its benefits and its dangers.
- Benefits from most types of exercise are not apparent immediately, and may take many months or years of sustained, regular adherence before significant and lasting results become clearly evident.
- Risk factors associated with lack of exercise do not materialize suddenly, but rather accumulate slowly over many months or years of inactivity, often in conjunction with overeating and other potential negative influences on health.

♦ Possible harmful effects from active exercise may take their toll suddenly and at any point (such as heart attack, torn ligaments, broken bones, or muscle damage), as well as gradually over time (such as bone spurs, stress fractures, or damage to joints).

As with global environmental issues, the decision whether to engage in a physical exercise program features problems and possible solutions that are mostly long-term, slow to change, and highly subtle. This makes it extremely, and frustratingly, difficult for us to see tangible short-term results from whatever course of action or inaction we follow. Instant gratification addicts are unlikely to be satisfied. Change happens, if at all, almost entirely imperceptibly and invisibly, and both improvement and deterioration take place with such glacier-like slowness that you would swear nothing is happening at all. Compared to this snooze-fest, watching grass grow is as breathtakingly fast-paced as a cross between NASCAR and ice hockey on crystal meth. And this apparent absence of movement, in turn, contributes immensely to the challenge facing the decision-maker—how to choose wisely in an environment of minuscule, protracted response to any stimulus. It takes so long to see visible evidence of change (whether good or bad) that many people just give up, and often never begin to get involved in the first place, under the assumption that there's either no problem or that nothing will change the status quo enough to make it worthwhile in any event.

If you thought the known factors were discouraging, you haven't seen anything yet. Well actually, you have, but what you're seeing are the knowns mired in waist-deep, super slow-motion mud. But it gets even worse when we turn to the unknown factors. What are the unknowns in the exercise-option equation? Let's list a few variables that most people would agree are relevant and appropriate to consider, yet resist precise definition.

- The probability that we have any latent medical condition (e.g., predisposition to aneurism, partial arterial blockage, etc.) that secretly renders us at grave risk under the heightened activity of vigorous exercise.
- The probability that any particular exercise regimen will actually trigger such a latent defect.
- The magnitude of harm we would suffer in such an event.
- The probability that our health or quality of life will deteriorate if we do nothing to change our exercise habits or lack thereof.
- The degree to which our health or quality of life will decline if we fail to exercise more.
- The rate at which such deterioration would happen, and the length of time it would take before serious symptoms become evident.
- The amount and type of exercise we would need to do in order to make a significant improvement in our health or quality of life.
- The length of time it would require before we would notice measurable improvement from whatever exercise program we pursue.
- The full extent of the benefits we would see, over sufficient time, if we diligently follow any particular exercise program.
- The added costs to us of engaging in a regular exercise program, in terms of money spent on equipment, athletic clothing, shoes, health club membership fees, pain relievers, muscle ointments, medical bills related to exercise-linked injuries, etc., and in terms of the time it would take away from our other activities (lost opportunity costs).
- The added costs to us of not exercising regularly, in terms of additional money spent on medicine and medical treatment, and income lost due to reduced productivity, work days missed, and diminished length of our active working life.

♦ The intangible losses from shortened and less enjoyable life, lessened ability to do the things we want to do, reduced energy and stamina, etc.

For anyone who has ever embarked upon an exercise program or, more likely, just thought about it, the similarities to our momentous environmental questions are painfully obvious. On a very basic level, we can argue that being overweight or otherwise out of shape is not something we've inflicted onto ourselves. Our could-be-better physical condition may in large part be attributable to genetics; maybe we were born with an innate propensity for obesity, a low metabolic rate, physical limitations on our ability to be active, etc. And if we didn't cause our problem, is there nonetheless a significant chance that we could substantially improve our situation through a strong, consistent exercise program, or is the problem really resistant to any such possibly-futile efforts regardless of how hard we try?

Moreover, a very formidable obstacle to our corrective action is the maddening gradualness of both the problem and the solution. We don't become overweight or poorly conditioned all at once, and we *definitely* don't lose large amounts of weight quickly, no matter what any glossy advertisements and hyper-enthusiastic infomercials claim. As with climate change and biodiversity loss, our physical condition declines very slowly, even imperceptibly, over long periods of time. We ourselves may not notice the extra 10 or 20 pounds we've gained during the past five years until and unless we discover that some of our seldom-worn clothes no longer fit, or we finally decide to get that physical exam and the doctor's scale gives us the bad news, or someone who hasn't seen us in several years shows up and comments (rudely!) on our recently-acquired excess baggage. That's the insidious, sneaky, wickedly subtle nature of such ultra-slow-moving accretions of infinitesimally tiny bits of bad news. We naturally tend not to be aware of the ever-so-slowly mounting hazard as it very gradually builds up over long expanses of time. Under those circumstances, we rarely if ever are struck with a sense of urgency about the problem, precisely because it sneaks up on us with all the rapidity of a sleepy sloth on tranquilizers.

Likewise, if we for some strange reason decide eventually to take action and do something about our slowly expanding equator, we don't get anything resembling instant gratification for our very immediate and palpable pains. It is evidently a universal constant that a pound we gained over one year will require three years of persistent, determined effort to lose. For vast, agonizing stretches of slow-motion time, our only apparent reward for all those sweaty, achy hours in the fitness center is the specter of our scale seemingly permanently frozen at the same weight, as a sort of permafrost monument to human futility. Even what feels to us like a full-scale, all-out, prolonged assault on the Battle of the Bulge, complete with regular, vigorous exercise and the best available diet plan, can be excruciatingly sluggish in producing any sizable and enduring progress toward our goals. This can be quite disheartening, and it makes it very difficult to muster the resolve to continue our program despite the disappointing short-term (that feels so much like long-term) results. Every instinct will urge us to give up the unpleasant, demanding, persistent self-denial and self-discipline our exercise regimen requires, because as far as we can tell none of it looks like it's worth all the effort and sacrifice. If we're talking about carrots and sticks, why should we continue to sweat and eat nothing but carrots when our excess fat still just sticks?

If this sounds familiar, it should help you to understand why problems like mass extinction and global climate change are so impervious to legal resolution. Unless we carefully find, and carefully examine, all of the correct relevant factors, and appropriately consider all applicable probabilities and uncertainties, the stark evidence before our eyes will argue powerfully and persuasively that we should do nothing different at all. Remember how fond we humans are of taking it easy on ourselves. For us, there's no action like no action. When it comes to hard work and genuine sacrifice on our part, when in doubt we're happy to leave it out. Our persistence is in our resistance. So we personally continue to gain weight, and our planet continues to lose species. It all evens out in some strange sort of way, right?

Well, no. But I hope this chapter has illuminated some of the ways we all struggle to make important life decisions without firm knowledge of all relevant factors. Even if we are unaware of it, we're all playing the odds when we choose whether to drive, smoke, exercise, have our kids inoculated, fly in an airplane, and countless other common personal dilemmas. There's no way around it. By necessity, all of us have to make choices—even if the choice is to make no changes—equipped with a dangerously incomplete and uncertain collection of pertinent information. We do it anyway, because we must. We might think of ourselves as non-gamblers, and we may pride ourselves on being so sensible, disciplined, in control of ourselves, and downright admirable. But on some level we *are* gambling, and not just on some penny-ante trivialities. We are gambling with our lives, as well as the lives and wellbeing of those closest to us. Now aren't you ashamed of yourself? Go get help.

So let's admit it. All of us gamble, and we gamble with some of the most incredibly important decisions any person ever makes. Once we acknowledge this, we are on the road to recovery, as individuals and as an entire species. But before we are ready to use this insight to make the right choices for our planet, let's see if we can gather additional tools by examining other forms of gambling as well. It couldn't hurt, right? The next chapter will consider how people use probabilities, incomplete information, and speculative cost/benefit analysis to decide how to deal with the classic forms of gambling—formal games of chance—sometimes with millions of dollars at stake. Let's go to Vegas, baby!

CHAPTER 9

WE'RE ALL ALL-IN

Let's move now to just a small selection of actual, no-kidding, literal forms of gambling with which many of us are familiar from our daily lives. Or maybe you've never gambled, but you have "a friend" who has, okay? Each of these games of chance has its own constellation of distinctive features, as well as some parallels to the more well-disguised gambling examples covered in Chapter 8. In a variety of ways, each type of formal gambling offers another view of how people often risk, usually lose, and sometimes win extraordinary amounts of money through ordinary encounters with uncertainty, gaps in important information, probabilities, and devilishly difficult dilemmas. I think it is useful to look under the surface at what's truly involved with some of these common and popular games of chance, and how we might apply some of the same principles in approaching the world's greatest environmental questions.

a. Lotteries.

Even if we don't live anywhere near a casino and would never think of wagering money in on-line betting, we are frequently reminded of the immense riches potentially awaiting us for the why-the-heck-not price of a lottery ticket. State governments trumpet the incredible payouts available from a cornucopia of lottery games, bombarding us with propaganda from giant billboards as well as advertisements on radio and television. The clear message is endlessly repeated: It could happen to you! You can't win if you don't play! Today's jackpot is at $165 million!

Depending on our personal circumstances, the decision whether to buy one or more lottery tickets at any given point in time may be easy or difficult, and a lark or a lifeline. Unlike the personal-life examples previously considered in Chapter 8, here we are literally facing a formalized game of chance, and the decision is limited to the narrow questions of whether to play at all, and if so, how much to bet. I think these are still in essence nothing more than variations on the theme we saw on display in Chapter 8—the dance of danger, doubt, and decision. These ever-present companions play and interplay all the time, for everyone, and with everyone.

This is so because all of human existence consists of our efforts to navigate our flimsy little life-raft with limited visibility and blinders on through hazardous rapids. But the formalized modes of gambling are worth our special attention, because they take place within a stripped-down, overtly probabilistic and artificial environment where risk and reward, odds and oddities, and incomplete information are present in their most potent and unadulterated form. Here, I believe we can see most clearly and directly how human beings grapple with mathematical truths and untruths, axioms and unknowns, certitudes and variables…with gargantuan stacks of money teeter-tottering in the balance. And this can be of value to us as we reach for the ultimate and optimal approach to "betting the Earth" in Chapter 10. So with this in mind, let's spell out some of the known facts about a typical lottery and see what the givens look like.

- ♦ A single lottery ticket costs very little, often only a dollar or so.
- ♦ We can purchase as many lottery tickets as we want on any particular occasion, as long as we have the money available at that time to pay for them.

- Many thousands, and even millions, of lottery tickets are sold for every occasion on which a jackpot is going to be available to be paid out.
- Depending on the type of lottery, there is generally only one very large payout possible per occasion, and sometimes a number of much smaller winners as well.
- Most lotteries don't guarantee that anyone will win the big jackpot on any particular day.
- Many consecutive days of play can pass with no one winning the top prize.
- As time goes by with no big winner, the size of the jackpot increases.
- As the size of the top jackpot increases, more people take notice and join the crowd buying tickets.
- The probability of any one ticket winning the ultimate jackpot is extremely low, given the vast number of tickets sold in pursuit of a single big prize.
- As any particular individual purchases additional tickets for any given lottery event, the probability of him or her acquiring the jackpot-winning ticket increases accordingly, ticket by ticket.
- Even if a person buys 1,000 tickets for one lottery event, he or she still has only a minuscule probability of actually capturing the big jackpot, given the millions of tickets often sold.
- The amount of the single large jackpot frequently reaches into multi-millions of dollars, many times more than the price of any one lottery ticket.
- The winning ticket in a lottery is determined by purely random chance, with no role for skill or expertise on the part of any person buying tickets (although some may disagree on this point).

- Lotteries determine winners on a regular, frequent basis, with new games and new tickets offered in rapid succession after the result of each jackpot drawing is determined.
- The cliché is true. If you don't play, you can't win.
- There is no way to be certain in advance what the result would be if you bet on the next game, or the next two, or the next thousand.
- There is no way to be certain in advance what the result would be if, for any specific lottery drawing, you buy an extra ticket, or an extra hundred, or an extra thousand.

Some devotees of lotteries may question whether there is in actuality no element of skill involved. There are people who insist that they can increase their chance of winning by methodically selecting the location from which they buy their tickets, or by using some system to bet on certain combinations of numbers (in games that allow such choices). But in my opinion these beliefs are based on anecdotal experience, a very limited and haphazardly selected sample size, an unrealistic or even primitively superstitious view of reality, and hearsay accounts of unverified events. I don't think they are grounded in objective, verifiable facts derived from rigorous experimental methods or controlled empirical observations. In making this threshold evidentiary ruling against admissibility of a skill factor in lotteries, I am illustrating another side to the issue I introduced in our discussion of inoculations in Chapter 8. Let's look a little closer and see what's going on here.

One of the great things about making decisions using my method is that we get to play the role of judge. We're the judge who decides the ultimate outcome of the question at hand, of course, but we're also the judge of what facts, figures, fables, and foibles warrant our consideration as "relevant and competent evidence." It's a key function of any judge to rule on disputes over what gets through the courtroom door as admissible evidence and what is excluded. The judge makes the call on what evidence passes the straight-face test and what fails to qualify for consideration under even minimal standards of relevance, credibility, and reliability.

But wait, there's more. We also get to play jury. The jury ordinarily decides how much *relative weight* to give to any particular item of evidence, once the judge rules that it is admissible at all. That is, the jury makes the decision, for each bit of admitted evidence, whether it should count for a lot, a little, or something in the middle as a piece in the big-picture puzzle. Some bits of evidence just have a greater throw-weight than others, and this too is a very important factor in determining how the overall question will be resolved. In my decision-making system, we, as the ones making the choice, must rule on both what is worth being considered in the first place *and* how heavily to rely on each specific item of admissible evidence we do consider. Hey, it's a lot of work, but would you rather trust it to some sticker-shock self-appointed expert to decide it for you?

I'll come back one more time to this threshold evidence-evaluation process in Chapter 10, because it is so central to my decisional system—or any other, for that matter. For now, suffice it to say that it changes the entire inquiry on lottery tickets if we rule out any role of skill in influencing our chances of winning. If we decide at the outset, as I have done, that we will not entertain even the possibility that skill is a factor, we automatically transform the nature of the debate in a deep and fundamental way. That choice makes a decision to wager money on lottery tickets seem much more frivolous and even irresponsible—kind of like risking money on the flip of a coin rather than a reasonable investment in our own skill and talent. It's a game changer.

I've presented a rather limited list of established facts relevant to lotteries, and I had to resort to playing judge and jury to include the absence of a skill factor among my roster of givens. On the other hand, there is no shortage of unknown quantities floating around in the mix of factors germane to the lottery-playing decision. Here are some of the more important variables or unknowns.

- ♦ The exact probability of any one ticket being chosen as the big winner in any particular lottery.
- ♦ The probability that we will ever win the big jackpot if we play the lottery to the maximum extent we desire for as long as we want.
- ♦ The number of dollars any specific jackpot would pay out in after-tax, real-purchasing-power terms.
- ♦ The amount of money we will wager on lottery tickets in any given period of time.
- ♦ The relation between the amount we will spend on lottery tickets and our fluctuating net income/outgo financial situation, i.e., the extent to which we will gamble less or more if we are in debt, at or below the poverty level, unable to afford the basic necessities of life, unable to pay our bills in full and on time, etc.
- ♦ What alternative uses we would have for the money we spend on lottery tickets, and which of those alternatives we would actually choose.
- ♦ The extent to which our expenditures on lottery tickets might cost ourselves and others essential or important goods and services that we could otherwise afford to buy.
- ♦ The degree to which it is important that we actually win a large jackpot or otherwise swiftly acquire a huge amount of money, and the comparative likelihood that any other means could supply the needed influx of funds.
- ♦ The extent to which winning the lottery would result in net positive impacts on our life, i.e., would do more good than harm.
- ♦ The mathematical relationship between the number of tickets we buy and the combined odds of our winning the big jackpot on any given occasion.

- Our ability and willingness to reduce or end our lottery gambling if it causes significant negative effects for us and those who depend on us for financial support.

- Our ability and willingness to reduce or end our lottery gambling if we lose a large amount of money and feel a strong psychological impulse to recoup those losses with a big win.

- The probability that playing the lottery at all will lead us to engage in harmful or compulsive gambling to a significant degree, whether in the lottery or other games of chance.

- The likelihood that we will consistently limit our lottery activities to a sufficiently low level of expenditures so as to cause minimal negative impact on our lives and those who rely on us for financial support.

The choice of whether to buy one or more lottery tickets is interesting: We theoretically could invest only the nominal amount necessary to buy just a single ticket (say, one dollar) and from that one ticket reap a fortune beyond the most fevered dreams of avarice. If we view the whole enterprise from that perspective, it seems crazy *not* to buy that one ticket. Just about anyone can find a dollar to spare, right? When we think about the Scrooge McDuck money bin heaped high with cash that could soon be in our backyard if we get lucky with the ticket that dollar will buy, the decision looks like a no-brainer. What if the next game coming up is the one we would win, if only we bought a ticket (or two, or five hundred)? Get out of my way, I'm in a hurry!

As I showed in Chapter 8, staying uninvolved brings its own special form of risk, just as the more active alternatives are risky in a different way. By not betting, we are taking a chance that we will miss out on what might become our one and only shot at hitting the jackpot. Isn't it also a gamble *not* to bet when that means we could lose our only opportunity to win wheelbarrows full of life-changing money? Yes, it is. There is a gamble that comes from betting and a gamble that comes from not betting. I will have much more to say about this concept and all that it implies in Chapter 10 as part of the discussion of what I call Type P and Type R errors. For now, it's enough to recognize that we are in fact gambling whether or not we actively buy a lottery ticket; it's just two very different types of gambling. So why not just take the active option and bet? The problem, of course, is that lottery tickets are like potato chips (or, for me, Cheetos). Nobody can stop at just one. Or maybe you are the one person who can?

Here's the most disturbing aspect of the lottery dilemma: Some of the key unknowns are things we don't know about *ourselves*, at the deepest, most personal level. All of us would like to think that we would never develop a gambling problem, and would never spend money on lottery tickets rather than food, rent, mortgage payments, insurance premiums, car maintenance, medicine, the electric bill, or shoes for our kids. But the sad truth is that some of us are very wrong about ourselves. We can never know for sure whether we would be able to stop when we've lost our 100th, or $1,000^{th}$, or $10,000^{th}$ dollar down the lottery's gaping jaws. We have no way of knowing whether we would resist the overwhelming urge not to quit while we're behind. We can't be certain that our occasional small wins won't be enough to entice us to keep playing "just a little longer" in pursuit of that "inevitable" big jackpot. We'll never know whether we might talk ourselves into cutting back on heat, fresh fruit, paying the rent on time, medications, new tires, life insurance, tuition, or a host of other alternative demands on our all-too-limited money in favor of investing another $50 this week to go after that all-time high jackpot.

The ratio of *potential* reward to *actual* risk from betting is overwhelmingly enormous in the individual one-ticket-purchased case. The huge mountain of money we *might* win just dominates the one dollar we *will* risk to give it a try. Because of this, and because the payoffs are both randomly determined and heavily publicized, the apparent soundness of a decision to play the lottery "just this once" is as seductive as it is illusory. As with slow-moving, imperceptibly incremental environmental problems like climate change or mass extinction, appearances can be very deceiving with lotteries too. It looks like such a nominal and safe investment, just purchasing one or two inexpensive lottery tickets when the jackpot is very rich. Why not bet, even if the odds are heavily stacked against us, when our failure to play could cost us the chance to win more money than we could ever earn in any other way for our entire life? But with such a low probability of winning, coupled with the seductive allure of a potentially enormous windfall win, we could end up throwing much more money away in the unending pursuit of that illusory jackpot, one ticket at a time. That is a very real risk inherent in all situations where we might invest in a low-probability but high-possible-reward proposition: throwing good money after bad, with no end in sight.

The "true" answer to the lottery issue, then, involves the resolution of some major unknowns. We'll never be able to nail down all those doubts before the fact, about ourselves or about the game we might play. So we decide—to play or not to play. And we decide in the midst of the mist, feeling our way through a thick and everlasting fog. We can only do this by making some assumptions—whether educated estimates or wild-ass guesses—about the value those key variables might actually have, even though we can never know for sure. And all the while, we are juggling all these Magic 8-balls with our personal life, our family's security, and the prospect of a windfall of wealth hanging in the balance.

b. Casino gambling other than poker.

Some of the same factors from the lottery question apply equally to our decision whether to wager money in a casino. It is invariably true that we can't win unless we bet, and by not betting we are always taking a chance on missing our golden opportunity to get rich quick. That's true in lottery games and all casino games of chance; we never know in advance what will happen if we bet the next round, or the next hundred rounds, but we do know that we have no chance at the big payoff if we remain inactive. One of the unknowns is always what will happen on the next dice roll, or wheel spin, or handle pull, or deal of the cards. The lottery mantra, "It could happen to you!" is fully applicable to every game of chance in every casino. Fueled by that intoxicating idea, immense amounts of money can change hands depending on who bets, who doesn't bet, how much the active gamblers are willing to risk, and the specific events on which they choose to invest their bets.

However, there are also significant differences between the two situations. Most of the ways in which casino gambling decisions diverge from the lottery situation stem from the far greater variety of games available in a casino, with many more types of bets available, each with its own probabilities and its own ratios of size of wager to size of payout. But different carefully calculated psychological pressures pull us toward wagering in each case as well.

In lottery games, usually only a few different choices are available, none of which (I argue) involve any skill on the player's part. All the options are just variations on the simple theme of betting money on some set of numbers being drawn by an independent stranger through a process of random chance. Sometimes we get to pick the numbers on our ticket, and sometimes we just get whatever is handed to us, but in either case, our influence over the matter ends with our choice of whether to play and, if so, how many tickets to buy. The odds against our buying the big-winner lottery ticket are invariably extremely long, but if lucky lotto lightning does happen to strike us (as opposed to the natural and decidedly unlucky brand of lightning that swiftly renders us into the form of a burnt king-sized tortilla chip) we know we will reap a gigantic payout relative to the small amount of our wager. It is the lure of that life-changing win that keeps the gigabets pouring into the system from hordes of hopeful, yet huddled, masses yearning to win big. It certainly isn't the excitement or glitz of the locations selling the lottery tickets, which often also deal in such glamorous commodities as toilet paper, gasoline, beef jerky, and petrified corn dogs.

The contrast with casino gaming is colossal. Especially in the jaw-dropping casinos on the Las Vegas Strip, but also in many other large casinos in Reno, Atlantic City, various Indian reservations, and numerous locations the world over, every possible stop has been pulled out to pull us away from our "stop" possibilities. Myriad colorful flashing lights inside and outside, towering dynamic electronic signs, ear-shakingly joyful noises from cascading avalanches of hard coins joining with thousands of cheerfully raucous bells, buzzers, and sirens, free alcoholic beverages in great abundance, plus many other even less subtle touches combine to create and maintain a relentless and electric atmosphere of high energy and excitement. It's a non-stop, over the top game of "Can you top this?" with one goal in mind: move the maximum amount of money from the sidelines into the game.

But there is also a vast contrast with lottery games in terms of the bewildering all-you-can-eat buffet options featuring many enticing and very different games of chance. These range from the impersonal, low-prestige, and mechanized varieties of slot machines available in endless profusion to the James Bond faux (or not so faux) opulence of the elegant baccarat rooms set apart from the crush and crowd of the main casino floor. These games may or may not feature a role for our skill in determining the probability of a win, and some incorporate skill much more than others, but none of them are particularly clear about what odds we actually face when deciding to bet on one thing or another. This is important, because in some casino games it matters a great deal which of the large and tempting spectrum of potential wagers we choose to bet our money on. The probabilities are far more unfavorable for some options than for others, but you wouldn't know it to look at the table in front of you. Warning labels don't look good in neon.

It is not my intention here to provide a complete gaming guide, although I am willing to consider any reasonable requests for lavishly-compensated individualized tutoring sessions at the finest Las Vegas establishments. For purposes of this book, it will suffice to outline some of the variations among casino games, and to point out how a logical decision whether and what to wager will incorporate some balanced treatment of the probabilities, alongside the plusses and minuses built into every gaming decision. With that in mind, here is a primer-level list of some of the (I hope) indisputable facts relevant to a casino gambling decision.

- Most casino games offer a minimum wager somewhere between $1 and $25 per bet, depending on the particular casino, day of the week, type of game, and the specific table.
- The maximum wager allowed per bet varies greatly, frequently extending into the thousands of dollars, and can often be negotiated upward.
- Many casino games proceed rapidly, with a new round of betting and payoffs every minute or two, depending on the particular game and the number of players.
- The amount paid per winning wager varies from game to game and from bet to bet within each game, with many paying off "even money" equal to the size of the wager and others paying many multiples of the wager's size.
- The facts, rules, and probabilities associated with each type of casino game are readily available to anyone who cares to learn about them, whether at no charge on the Internet or a free table-side handout, or from a huge number of affordable gaming guides and how-to books.
- Unlike lotteries, and except for a few slot machines, there are no casino games that offer million-dollar winnings from any single play. Wins and losses in the six- or seven-figure range are ordinarily accumulated only over the span of many plays and over a comparatively extended period of time.

- Casino games generally place a maximum and minimum on the amount any player can wager on any particular event, which in turn determines the range of possible wins and losses on any particular roll of the dice, spin of the wheel, pull of the slot machine handle, or other gaming event.

- The maximum and minimum wager limits apply only for any one specific event (spin, roll, hand, etc.), and apply anew for each subsequent event, so that over a span of time a player can, in the aggregate, risk many times more than the maximum amount for any single event.

- Some people are able to limit their gambling effectively and can quit whenever their losses reach a certain predetermined limit; others cannot.

- No betting system or method of varying the size of wagers, depending on such phenomena as the number of times a particular outcome has occurred consecutively, can reduce or eliminate the mathematical advantage the casino enjoys over the player.

- The "house percentage," which represents the mathematical advantage the casino has over every player, differs enormously from game to game, often with great variation on particular bets within each game. Some common examples follow.

- At American-style roulette, where there is both a single- and double-zero on the wheel, the house percentage is a constant 5.26% on essentially all betting options. This means, on average, for every $100 we wager, we "win" back only $94.74 and lose $5.26.

- For baccarat, betting on the "player" has a house percentage of 1.36%, while bets on the "banker" feature a house percentage of 1.17%.

- Casino craps (dice) has a wide array of over 100 different available bets with tremendous variation in disadvantage for the player. The highest expected-value wagers are "pass/don't pass" and "come/don't come" with full "odds" taken, with a relatively low house percentage of between 0.6 and 0.8%. "Field" bets are much worse, with a 5.5% house percentage, and it goes downhill from there, all the way to 13.89% on "any two or twelve" and 16.67% on "any seven".

- Slot machines vary wildly, depending on many factors, and house percentages often range from about 1% to 5%.

- Blackjack is a complex game featuring a major component of skill on the part of the player, and skill exerts a significant effect on house percentage. The number of decks of cards used, together with variations in rules (such as whether the dealer will hit or stand on "soft 17") and playing procedures also affect house percentage.

- Depending on player skill, the way the cards are shuffled, and these other factors, house percentage for blackjack can range from a small but mathematically significant advantage for the player (where the player is adept at an effective "card counting" system and is able actually to use it under favorable playing conditions) to a house advantage among the worst of any casino games.

- Casinos actively employ a variety of counter-counter-measures to prevent skillful card counters from regularly winning money at blackjack. These measures include barring successful counters from playing, shuffling the cards frequently or even perpetually, not dealing deep into the deck(s), and using several decks of cards mixed together.

- For any casino game, the player's discipline with money management is a very important factor. Players who chase their losses in a desperate attempt to catch up, increasing the amounts they bet and refusing to quit when behind, are likely to make their losses far worse.

- People who gamble with money they objectively cannot afford to lose are at risk for jeopardizing their ability to pay their bills and meet other financial responsibilities.
- Some people become compulsive gamblers and regularly bet, and lose, money they need for life's necessities.
- People who never bet on any given casino game have zero mathematical probability of actually winning any money from that game, whether in small, moderate, or enormous amounts.
- Some people at least occasionally win large amounts of money at casino games, even many thousands or millions of dollars.
- When people gamble while excessively tired, or hungry, or intoxicated, or emotionally upset, their judgment tends to be less sound than normal, all else being equal.
- When playing any casino game or making any bet in which the house has a mathematical advantage, probabilities favor the casino eventually taking most or all of any given player's money. Under these circumstances, the more a player wagers, and the longer he or she plays, the greater the likelihood the casino will take the player's money.

With these rather sad facts of life established about casino gambling, we are prepared to take a hard look at some of the unknowns that riddle the riddle of the casino's vault. The following is a litany of a few of the more significant variables that cloud our dilemma of whether, how, and how much to gamble. Remember that casinos are professionally designed and actively managed to render these variables as unclear and well hidden as possible. That camouflage is vital to the perpetuation of the casino party. Someone, after all, has to pay for all those pretty lights. So what is it that we should know that we don't know?

- The probability that any given gambling session, stretching over hours or days of extended play, will bring a net gain or loss of life-changing amounts of money.
- The likelihood that we will have the resolve to stop playing if our losses from any given session exceed our pre-determined limit or some other reasonably affordable level.
- The chance that our luck will change significantly for the better if we continue to bet after losing a particular amount of money.
- Whether we will recoup all of our previous losses and then some if we place a large enough wager on one more play, or several more plays, or many more plays.
- The extent to which we will have sufficient self-control to quit gambling if our losses erode our ability to meet life's responsibilities.
- Whether we will consistently persist in playing skillfully and prudently if we suffer one or more large, unfortunate losses, or whether we would lose focus and gamble recklessly in a hasty attempt to recover.
- The chance that we will restrict our betting to only those options that offer players the highest Expected Value while avoiding all of the low Expected Value games and wagers, even when we have lost a large amount and feel the need to get even.
- The probability that our play would be significantly undermined by distractions caused by any combination or aggregate effect of our fellow gamers, dealers, casino executives, bystanders, nearby sexual temptations, loud noise, bright flashing lights, crowds, or other environmental influences.
- The extent to which we will succeed in quitting while we're ahead if we win a reasonably large amount, or whether we will continue gambling until we give our winnings back.

- ◆ The specific degree to which our ingestion of a given quantity of alcoholic beverages will degrade our skill, discipline, judgment, and emotional equilibrium while gambling.
- ◆ Whether we will possess the requisite self-awareness to know when to quit if our fatigue, hunger, psychological state, health, sobriety, or other factors deteriorate to a level that makes continued gaming unadvisable.
- ◆ The probability that we would deal effectively with weeks, months, or even years of sustained gambling losses and confine the damage to tolerable limits.
- ◆ The probability that we would react emotionally and irrationally to any particular gambling setbacks in such a way that would lead to reckless, mathematically unsound, undisciplined, and irresponsible playing and wagering decisions.
- ◆ The likelihood that the time, money, and effort we expend on gaming would cost us and those who depend on us more in both tangible and intangible terms than the rewards warrant.

This complex tangle of facts, fallacies, and fractions adds up to a fog that would send Heisenberg himself dashing toward Dr. Freud's couch. Much of this, of course, is (rather ironically) most definitely *not* due to chance, but instead is the desired outcome from a huge amount of strategic planning and meticulous design. The one thing about casino gambling that does not depend on random probabilities is the psychological care invested in creating a powerful push to bet, bet big, and bet often. Even the lowest Expected Value wagers available (such as "any craps" at the dice tables) are lovingly designed and dressed up to *seem* like a worthwhile place to invest a pile of our money. There certainly are no warnings posted alerting us that these are options where the house percentage climbs into nose-bleed heights. The casino employees will not try to dissuade us from risking our resources on bets where we effectively give up on average one-sixth of our wager on every play. On the contrary, sucker bets like the "propositions" on craps are all tricked out, front and center, with big, colorful numbers and letters and the promise of a hefty payout if we should happen to hit a lucky outcome. Dozens of choices like these are professionally disguised by talented, Hollywood-ready makeup and special-effects specialists to look like great investment opportunities instead of a bunch of little lotteries just waiting to piranha our paychecks. As long as we don't look too closely, we'll never notice all the cosmetics and costumes.

The sneaky natural slowness and subtlety of global environmental threats has a flashy first-cousin in the artificially-crafted but equally stealthy mirage of casino gaming. In both cases, whether by the influence of nature or by human purpose, the true situation is difficult to see through all the fog. Major risks are camouflaged to appear normal, non-threatening, or even good news. A convoluted Gordian (or should that be Gödelian?) knot of interlaced unknowns and hazards is decorated to resemble a big birthday present tricked-out with extra-generous amounts of pretty ribbon and gift-wrapping. Considerable dangers are concealed because they happen little by little and take a long time to exert a perceptible toll, whereas the benefits of business-as-usual are obvious and immediate. Minuscule, numerous, gradually accumulating incremental costs associated with the hidden threats are easily overwhelmed, from our psychological perspective, by the much larger instantaneous benefits offered by the alternative, status-quo course of action. But conversely, both in the casinos and when "betting the Earth," we also are gambling when we *refuse* to place any wagers at all. When we choose not to risk any of our money, or decide not to wager enough to make the big prize achievable, we are actually accepting the risk that we will miss participating in one or more pivotal situations. Again, as I will explain in Chapter 10, there is a genuine risk inherent in inaction—risks I am going to call by the mysterious names of Type P and Type R. So stay awake until we get there. Unavoidably, the peril buried inside inertia is the risk of sitting out the event that could have changed the world, or won life-changing avalanches of money for us and our families, if only we had gotten involved and invested sufficient resources to make our big game-changing win possible. That, too, can reasonably be viewed as a loss, and a gigantic one at that, despite the apparent—and illusory—absence of risk-taking action inherent in deciding to remain out of the game and above the fray.

Under the circumstances, once we step back and take in the entire picture it's easy to see why so many people are deceived by the similarly insidious wiles of casino gambling and global environmental degradation. Our normal frame of reference tends to be focused on the readily-apparent, here-and-now issues that stand up tall and shout out for our attention, because many times the obvious problems really are the most urgent and important. The difficulty lies with the fact that once in a while the image before our eyes lies. Unless our decisional methods check for slow, deceptive, baby-step, well-concealed hazards as well the ones that rush right up to us and scream in our face, we can be taken by surprise and remain on auto-pilot until it's too late to turn around.

Casinos depend on this. All the loudest, brightest, largest, and most blatant aspects of the casino environment are specifically designed to foster an atmosphere of excitement, fun, carefree indulgence, easy wins, free-flowing money, and time-stands-still partying. Casinos are deliberately and very calculatedly organized to generate irresistible fear and anxiety in all of us that we will lose our only chance to win a huge payoff if we keep our wallets closed! It takes a powerful and determined focus on the far quieter and more obscure dangers to keep from being swept along with the nearly overwhelming addled-adult-adolescent, hormone-overloaded riptide. Unless we take the trouble to make ourselves aware of the trouble, and then actively search for evidence of danger, we are at great risk of being caught by surprise. Even though all the pertinent cautionary information is easily accessible by anyone who looks for it in the right places, the casino environment itself is deliberately crafted to discourage any thought of a potential downside to all the dizzying thrills. There are certainly no immense, flashy, raucous, prominently located displays loaded with thoughtful books about the relentlessly unforgiving house percentages. If you want the whole truth and not just the "You can't win if you don't bet!" message, you must affirmatively go out and look for it.

c. Poker.

I want to use poker as one final example of gambling decisions in my search for illuminating parallels to the environmental dilemmas staring at us. Even though poker is often played in the same brick-and-mortar casinos as the games we discussed in the previous section (blackjack, roulette, craps, and the like), many people also "play" poker on the Internet for real money. This distinction, coupled with some other unique aspects of poker, makes it different enough to warrant separate consideration in its own right.

There are many varieties of poker, including Omaha, Razz, and Stud, and some varieties also include a number of sub-variations as well. However, for purposes of this book, I'm going to look at only one type of poker, usually called No-Limit Texas Hold 'em, or just "Hold 'em" or "NLH" for short. During recent years, Hold 'em has emerged as the overwhelming favorite of poker players all over the world, and it has become difficult even to find other poker varieties offered in many casinos and Internet gaming sites. The prestigious annual Main Event of the World Series of Poker in Las Vegas, and its newer cousin in Europe, both use Hold 'em as the only game for all the competitors. For these reasons, we'll confine our analysis specifically to the Hold 'em variety of poker.

If you are trying to decide whether to play poker, and if so, how to go about it, you need to deal with many of the same issues that pertain to all forms of gambling, but you also must grapple with some major departures from that norm. For example, in poker (unlike any of the casino games from the previous section) players genuinely compete against each other, not just against the house. Because of this, there really is no house percentage advantage on the side of the casino (although casinos generally charge a fee or "rake" for the privilege of playing, as a percentage of the amount bet), but rather each player's probability of success is linked to the quality, quantity, and luck of the other players. And, as with blackjack, poker incorporates a significant element of skill as opposed to pure random chance in determining who wins and who loses, and in what amount. The skill component of each player relative to the other players becomes increasingly important as the duration of competition extends from just a few quick hands to long, sustained interaction over extended periods. So, unlike the situation with, say, roulette or slots or baccarat, a poker player is at least in part the master of her own destiny and not just a helpless captive of the whims of chance. A poker player's skills can be honed over time through study, experience, observation, and sharing of ideas; and this can reasonably be expected to pay dividends eventually in the form of better results at the table (or the computer).

One of the hallmarks of No-Limit Texas Hold 'em poker, as the name itself implies, is that there is *"no limit"* to the amount we can wager on any given pot, or at any stage of a given pot. Unlike all other casino games, with poker there is no maximum ceiling limiting our bet size. The only limit is the amount of chips we own. That means that on any hand, we are free to place *all* of our money at risk, and to do so at several junctures: before the flop, after the flop, after the turn, and after the river card is dealt. That adds up to an over-the-top, non-stop profusion of opportunities to bet everything we have! And, depending on the relative size of our stack of chips compared to those of our fellow players, and depending on how many other players choose to join us in risking their chips, we therefore have numerous chances to double, triple, or otherwise multiply our chip stack, pot after pot. We also have numerous chances to lose every last chip we own, just as abruptly. The boom—and bust—possibilities come fast and furious, and players come—and go—from the game just as fast and furious. I've personally noticed that players tend to leave the game considerably more furious than they were when they arrived. Go figure.

Once more, we'll list some rather uncontroversial facts about poker. These points appear to be relevant to the decisions anyone faces when contemplating getting into the game. I've tried to limit this litany to facts that most knowledgeable people would agree are objectively true rather than mere matters of opinion (and there are plenty of things about poker that are very much matters of opinion). If I've failed, well that's poker.

- ♦ Luck, or random chance, plays some role in determining success or failure at poker.
- ♦ The skill of each individual poker player also has an influence in determining his or her results.
- ♦ The skill of any given player in combination with the skill of each of his or her opponents, as a relative comparison, exerts an influence on each player's success or failure.
- ♦ Player skill generally has less opportunity to drive the outcome of poker during brief periods of competition than over days, weeks, months, or years of play.
- ♦ Irrespective of individual or relative player skill, even at the highest levels, luck is always a factor in determining poker results.
- ♦ There are defined, well-understood rules to the game, including the relative rankings of the various possible hands (e.g., a flush beats a straight, quad kings beats quad queens, any two pair beats any one pair, even aces, etc.).

- Because of the way poker is structured, there are definite, precise, and readily available probabilities of any player being dealt any two starting pocket cards (e.g., 220-to-1 against being dealt pocket aces).
- Other mathematically verifiable probabilities are inherent in the game of poker, such as the odds that any two specific starting cards will win a pot against a random selection of other starting cards.
- Even a hand consisting of the strongest possible two starting cards (a pair of aces) is not mathematically guaranteed to win a pot against even the weakest possible two starting cards (a seven and deuce of different suits).
- As a given pot is played out and more "community cards" are dealt, mathematically verifiable probabilities apply to each specific hand of three, four, or five cards and its chances of winning against random opponents.
- There are "tells" or clues that a skillful player can use to try to deduce to some extent the range of hands the other players might have, at least in face-to-face competition.
- Under some circumstances, some players are very adept at inferring the relative strengths of the other players' hands, through reliance on physical tells, betting patterns, and past behavior.
- Some players are better than others at concealing their emotions and intentions and minimizing the number of tells they give off.
- Some players, at least to discerning opponents, are relatively easy to "read" due to their ineffectiveness at eliminating their physical tells, the predictability of their betting patterns, and the predictability of the manner in which they play particular hands.

- If a player bets enough or in such a way as to induce all the other players at the table to fold their hands, he or she can win a pot with even an extremely weak hand.
- If a player's bluff doesn't succeed in driving everyone else to fold, the bluffer is often at a significant statistical disadvantage against one or more stronger hands that remain active.
- In NLH poker, even a huge quantity of chips can be lost quickly.
- If a player has any chips at all left and is still in the game ("a chip and a chair") there is still some possibility of recovering and eventually winning all the chips in the game.
- In any given pot, the more chips you bet, the more chips you can potentially win or lose.
- If you push all of your chips into the pot (move all-in), you are placing yourself at risk of going broke, in the event you lose to another player with a better hand and an equal or greater stack of chips.
- In poker tournaments, unless players are expressly allowed by the rules to re-buy more chips, any given player is completely and permanently eliminated from the competition once his or her last chip is lost.
- A poker player's physical, mental, and emotional condition can exert an influence over the results from the competition.
- If a player goes "on tilt" and becomes angry and emotionally upset because of what he or she perceives as bad luck, obnoxious behavior, unfair treatment, or other distractions, the quality of that player's decision-making is likely to be adversely affected.

Let's stop at this juncture, since this isn't primarily a book about poker. Take another look at all of these reasonably well-established givens that are relevant to our success or failure at poker. It's an impressive list, and we could easily add to it. We know, and know very well, quite a lot of important information about poker. But now reflect on all the salient features of a poker game that are *not* sure things, and how much uncertainty skulks around the ankles of even the finest, most experienced poker players in the world. Just ask Phil Hellmuth, who famously declared after being eliminated from a poker tournament, "If luck weren't involved, I guess I'd win every one."

♦ Under most circumstances, and depending on the combination of community cards on the board, even a very strong hand (such as top two pair, or three of a kind) is not absolutely certain to win a given pot.

♦ With some boards (such as multiple cards all of the same suit), even extremely strong hands such as a straight or a flush are far from guaranteed to win the pot.

♦ Depending on the cards on the board, sometimes even very rare and exceedingly powerful hands such as quad aces or a straight flush can still lose to an even better hand (e.g., a higher straight flush or a royal flush).

♦ The best possible hand ("the nuts") pre-flop is pocket aces, but at every subsequent stage of the hand (the flop, the turn, and the river) the nuts can suddenly change in dramatic and unpredictable ways.

♦ Until and unless the other players' hole cards are revealed, any given player can never know to 100 percent certainty what is in any other player's hand.

♦ Depending on a combination of circumstances, a large wager by an opposing player may signify anything from a very strong hand to a very weak hand (a bluff).

♦ Conversely, a small bet (whether a minimum-size raise, or just a call of the other players' bets) can signify anything from genuine weakness to a monster hand that the player is slow-playing to lure others into betting and thereby enlarging the pot.

- After the flop, a player who just "checks" instead of betting may be attempting to conceal a monster hand or may in fact actually have a fairly weak hand that doesn't warrant risking more chips.

- Regardless of how well we have studied a given player's methods and behaviors, there is always the possibility that he or she will do something unexpected at any stage of any particular hand.

- Even if we can't imagine anyone making a particular move (because it has an extremely low mathematical probability of success compared to the large risk involved), there remains some possibility that one of our opponents will in fact make that move at any time.

- If we risk all of our chips on the assumption that the only other player remaining in the pot never would have called our very large raise with a certain specific weak hand (e.g., seven-deuce off-suit, or eight-trey off-suit), there is a low but non-zero mathematical probability that our opponent will turn out to have done exactly that.

- Under circumstances where only one specific set of pocket cards could combine with the board to beat us, irrespective of how stupid we may think it would be for anyone in that situation to raise, re-raise, or call a raise with those cards against us, there is some non-zero statistical chance that our opponent has in fact been exactly that stupid, and will be exactly lucky enough to win all of our chips.

- Irrespective of how many consecutive hands have resulted in what we consider extraordinarily bad luck for us and/or good luck for our opponents, there is some real, non-zero probability that this unfortunate trend will continue for the next hand and beyond, even sickeningly well beyond.

- Although we know that, statistically, we have a 1 in 221 chance of being dealt pocket aces on any given hand, we have no guarantee of actually receiving those cards even once over extended periods of play involving thousands of hands.
- Conversely, although any one of our individual opponents also has only a 1 in 221 chance of being dealt pocket aces on any given hand, we can never be certain that he or she will not receive pocket rockets on three or four successive hands.
- No matter how many chips we bet on any given hand in an attempt to get the other players to fold, we can't be sure that every player will in fact fold, even if some of them have very weak hole cards.
- Even when you finally get a hand where you have "the nuts" and can't lose, and regardless of how you behave or how much/how little you bet, there is no assurance that any of the other players at the table will wager any chips and enable you to win their money.
- We may have an audacious opponent who moves all-in before the flop on twenty consecutive hands, and wins every pot uncontested with what probably included several bluffs on very weak hands, but on the twenty-first successive all-in we can't be certain of winning if we call their all-in with all of our own chips, even if we hold pocket aces.
- Even the most successful and highly skilled poker player in the world has some chance of making careless, novice-level mistakes, such as misreading his or her own hole cards and misplaying the hand as a result.

Anyone who has ever played poker will find much that is both familiar and disturbing in this list. I've personally competed in the World Series of Poker Main Event at the Rio in Las Vegas more than once, and I can testify from sad experience how even the greatest players in the world suffer at the "hands" of Bad Beat, the god of Poker Probabilities. Several entire days filled with twelve-hour sessions of intense concentration can be undone in a single instant of unlikely and unforeseen misfortune. You can perform with great control and discipline over the span of hundreds of hands of fierce competition, only to see all your carefully-amassed chips vanish in one pot where you could only lose if one card in the deck came out on the river—and it did. If you don't know how vast the chasm is between 95 percent and 100 percent, ask a former world champion who pushed all-in with only a 5 percent chance of losing the pot—and lost it all. And if you were wondering what uncertainty feels like in your gut, try placing a very large bet before the flop when you're holding pocket kings, and finding yourself called by a player who then moves all-in when an ace shows up on the flop. Poker players have a very fitting one-word name for that sensation: sick.

In the highest of high-stakes poker tournaments, players must quickly make game-changing and life-changing decisions based on incomplete and uncertain information. Whoever wins the annual Main Event captures a cash prize in the amount of several million dollars—ungodly bundles of money for any mortal and for a few demi-gods as well. These tournaments may not literally be a matter of life and death, but the results can nonetheless make an enormous difference for the winner and his or her family for many years to come. With once-in-several-lifetimes money at stake, and a family full of people depending on you, how would you decide whether to call an opponent's huge raise that would require you to risk all of your remaining chips? Or, facing a large raise from a player to your right with a giant stack of chips, how would you make the call whether to make the call? When we're watching a tournament on television at home, these choices seem easy and obvious, and we effortless shout out, "Go for it!" and tell ourselves we wouldn't hesitate if we were in that player's chair. Take it from me, though, those dilemmas have much sharper horns when it's your own future sitting there in front of you in the form of your chip-chunked skyscrapers.

Remember, you'll very rarely have the absolute "nuts" in any of these tipping-point defining moments, so there will almost always be at least some small probability that you will lose if you commit all your chips to the pot. And in tournaments like these, lose means lose...once your last chip goes bye-bye, you go home, with no second chances, no re-buys, and no do-overs. It's strange but not funny how suddenly your life-changing money can get shipped off to change someone else's life instead. Poker chips can be so fickle. So considering how much is riding on your doubt-drenched decision, what level of likelihood would *you* require before you risk everything on a single hand?

There are deep layers of uncertainty inherent in such decisions. Before you choose whether to commit your chips to the pot, you don't get to see the other players' cards, and this obviously creates a considerable degree of doubt in the mind of any player, even a highly-experienced professional who takes pride in being quite adept at estimating other players' hole cards based on their tells. But even if we could see our opponents' cards, which we can't, the result of this x-ray vision would almost invariably reveal some questions as to the ultimate outcome of the hand up until the moment when the last community card is dealt on the river. Our hypothetical Superman-like vision might show that there is only one single, very specific, card still remaining in the deck that could come out on the river and cause our ace-high flush to lose to a full house, but that lone card would cost us everything.

Moreover, if there are other opponents left to act after us, we can't know for certain whether they will fold or likewise move all-in if we push all of our chips into the middle of the table. The possible addition of more opponents who stay in the hand can open up more ways in which we could lose the pot depending on what community cards are dealt, so this too is an element of unknown facing a poker player. Generally, if you don't have the absolute nuts with victory a mathematical certainty, it's riskier to face more rather than fewer opponents in a given pot, because there are greater possibilities for unusual card combinations to materialize and defeat your hand.

In poker, as with the global environmental dilemmas, much depends on the extent to which the decision-maker is willing to accept risk of ruin. Some poker players are famous for being very aggressive and willing to bet most or all of their chips to win large pots. Because of their readiness to invest large amounts, they can often bluff more cautious opponents into folding better hands. The price these aggressive players pay is the increased risk that opponents will be willing to call their bluff in the hopes of catching them with a weaker hand. Implicit in the aggressive players' approach is an acceptance of facing an occasional huge pot where they are either in a coin-flip situation or even at a significant mathematical disadvantage.

But even some bold gamblers become more careful when friends and loved ones are depending on them and they are in a position where it is important to survive deep into a tournament. That scenario more closely resembles the "betting the Earth" circumstances, where far more than our own individual, narrow self-interest is at stake. When we have people we care about relying on us to make correct choices on their behalf, our eagerness to take sizable risks generally decreases, and can even approach zero where survival is considered paramount, and the prospect of ruination is too unacceptable to warrant assuming so much as a small probability of disaster.

Conversely, in poker as in life, there is literally no risk-free option. To keep pace with progressively larger blinds and antes, even the most risk-averse player must take some chances, and be willing to invest enough chips to fend off bullying bluffers or aggressively solid players under the right circumstances. Even with considerably less than a mathematical certainty of winning, every player has to pick the right spots to invest most or all of his or her stack, to have any chance of surviving deep enough into the tournament to win some money. Under rather strict time constraints, and with *far* from perfect knowledge as to which community cards will be dealt, what other opportunities might come on subsequent hands, and all the opponents' hole cards, playing patterns, and degree of risk acceptance, each player is forced to decide, and decide repeatedly. Fold or bet? Call a big bet or fold? Check or bet? Call or raise? Raise a little or an average amount? Raise an average amount or raise all-in? Even a single big pot can require a player to make these decisions multiple times (pre-flop, flop, turn, and river), with further possibilities for re-raising, re-re-raising, etc., at any stage. The trouble is that we never know in advance which handful of pots out of the hundreds or thousands we see during a long tournament will be the pivotal ones that change everything. But it is precisely those very rare but crucial big pots that determine, to a very large extent, whether our tournament terminates prematurely or extends long enough to pay major rewards.

It is generally accepted to be true that poker players in a tournament should fold considerably more hands, without even betting, than they play. Because we usually are dealt low-value hole cards much more often than large pocket pairs or other promising starting cards, and because every time we invest in a pot there is a chance we will get caught up in a raising war that costs us our entire stack, we should just fold most of the time. It's safe, it's smart, and it reflects the probabilities. Much more often than not, it simply isn't worth it, in terms of risks and rewards and mathematics, to invest anything at all in a speculative situation. Thus a wise poker player will spend most of the tournament not playing hands at all. It's boring, but it's the best alternative most of the time.

This is arguably an example of the Minimax method in action, which we introduced briefly in Chapter 4. If the worst-case scenario for a poker player is considered to be the situation in which she loses all of her chips in any single hand, then Minimax would nudge her in the direction of avoiding the all-in bet. She would try to play only small pots, "chopping" at most hands with moderate wagers in an attempt to stay away from losing her entire stack. There is room for judgment here, because a lot depends on how strictly we want to eliminate any possibility of the worst-case scenario. Are we satisfied if we "minimize" that possibility to a low but non-zero probability, such as a five percent chance of ruin, to open the door to potentially huge wins? If so, we might push all-in when we believe our pocket Aces are up against a lower pocket pair and, at the turn, our opponent has only one more chance to draw a third card for a winning "set". We have a 95 percent chance of doubling our stack if we move all-in and our opponent calls, in that scenario…but there is that 5 percent probability that they will river the dreaded set and take all our chips.

One way of approaching this situation under Minimax would be to refuse to risk all of our chips on any pot where we don't have "the nuts" and thus lack a mathematically certain victory. With this application of Minimax, we would literally minimize (to zero) the maximum negative outcome (losing all our chips) on any particular hand. However, such an extreme risk-averse approach would make us vulnerable to a slower and more insidious worst-case scenario in which we play so cautiously that we eventually, and very gradually, lose our entire stack to the ever-escalating blinds and antes. That is an equally horrible outcome in which we end up out of the game, but it happens a little at a time as we refuse to take the calculated risks that might give us enough resources to stay alive. So, depending on how we look at the worst-case scenario (sudden versus incremental loss of all our chips), Minimax might indicate the appropriateness of an occasional very large wager, given sufficiently favorable odds.

This is an important point. Even though our decision analysis tells us that we need to stay safely out of the action most of the time, there's still that 20- to 30-percent of the pots where we can and should get involved…and indeed, it would ultimately be a disastrous gamble *not* to gamble. A "rock" who just sits out too many pots will never get enough premium hands to keep pace with the ever-escalating blinds and antes. They just aren't plentiful enough. The small but relentless accumulation of maintenance costs and lost opportunities gradually takes its terrible toll, like the "death by a thousand cuts," and the final result is ruin. It is slow, incremental, agonizingly protracted ruin, but ruin nonetheless, just as bad in its ultimate impact as the variety that crashes down on us all at once like a guillotine's blade. Again, it's a matter of steering between the Scylla and Charybdis—or between the death by a thousand cuts and the death by guillotine—of the two dangers. We gamble when we bet, and we gamble when we don't bet, too.

All the while the tournament is going on, a poker player's environment is rapidly evolving, forcing him or her to make these life-changing decisions within a matrix of ever-shifting knowns and unknowns and ever-modulating probabilities. This is true from hand to hand, and from one stage of any given hand to its next stage. Does a (likely) 80 percent probability of winning a pot with at least $10,000 worth of chips in it warrant betting the last $5,000 of our chips? If our bet might (with some undetermined but non-zero probability) entice another player to enter the pot and improve our implied odds, does that change our decision? If we've already invested half of our remaining chips in a pot, and our opponent makes a huge re-raise that would put us all-in if we call, do we conclude we are probably beaten and it isn't worth jeopardizing the rest of our stack to find out, or do we press on and let the chips fall where they may? This happens with some regularity, where our opponents' betting patterns might cause us to reassess our early assumption as to the strength of their hands. It could force us to accept the possibility that an opposing player has one of the very few sets of hole cards that can beat our own powerful hand.

In fact, one of the most dangerous situations in poker, and in our environmental dilemmas, relates to this situation where we're tempted to risk everything to take advantage of an opportunity that is almost, but not quite, a sure thing. This deadly trap is the exact opposite of the proverbial blessing in disguise. It's damnation disguised as a blessing. It happens where our hand, within the context of the community cards, is within a whisker of being mathematically certain to win the pot, and someone else is (foolishly, it seems) willing to invest a massive number of chips to beat us. Our quad kings, for example, on a given board could only possibly be topped by a straight flush completed by a wildly improbable couple of hole cards. We might think that no marginally sane opponent would have bet, raised, and re-raised so much money early in the hand with the only hole cards that, after the turn and the river, just might have ended up as the missing fill-in pieces of that straight flush. It's almost irresistible to bet all we have under these circumstances. And yet there's always that slim but terrible chance that our opponent was in fact either insane, stoned, or breathtakingly aggressive…and has just gotten luckier than anyone has any right to be. At our expense.

These damnations in disguise can cost us incalculable losses in poker. Likewise, in our environmental decisions, we need to think carefully and creatively about every uncertainty facing us, and about whether a slim risk of unfathomable loss is worth the gain we might attain if the situation plays out in our favor. Both global climate change and a modern mass extinction entail the specter of catastrophic environmental harm if we make the wrong decision, including the decision to take no action when in fact action was demanded. But they also could consume dizzying quantities of our scarce resources (money, foregone opportunities, effort, quality of life, etc.) as we mount a Marshall Plan type of all-out counteroffensive, which might turn out to have been unwarranted. That's what happens when you are forced to decide the highest of high-stakes issues without first seeing all the cards face-up on the table. That's the life game we call poker, and that's the poker game we call life.

Poker decision-making is a constantly moving target, and it likes to wear clever camouflage and lob lots of smoke-bombs, too. Rational, intelligent people know the importance of gathering reliable evidence and then assessing it in conjunction with prudent risk/reward analysis, with the proper relative weight for each relevant factor in light of the applicable probabilities. So far, this sounds like a difficult but not impossible question from an upper-level math exam. But in reality, poker requires you to do all of this with appropriate adjustment for major uncertainties on probabilities, and even on underlying concealed facts...on multiple levels, and at multiple stages during any given hand. Once you grasp this, the challenge looks more like a graduate-level final exam in advanced chaos math or quantum mechanics where we forgot to drop the course and haven't studied. That's one of my worst recurring nightmares, but it's also fully-awake, everyday real life for poker players. It's enough to give me a drinking problem to go with my gambling problem and my cheesecake problem. A trifecta of compulsions.

One of the strongest parallels between the decisional process in poker and in global environmental questions stems from this: the long lag time between cause and effect for some aspects of both problem and solution. In poker, we may not realize that our flawed play of certain hands, our tells, and our suboptimal betting methods are undermining our long-term results. Poker, like any system that depends heavily on probabilities, can be very slow to reveal in unmistakable clarity the results of small differences in advantage and disadvantage from particular tactical moves. If, for example, a player gains an extra two percent overall mathematical advantage by betting rather than calling in a given situation, that marginal difference is small enough that the results may be hidden from an ordinary observer—maybe for a long time, maybe forever. Unless a player keeps meticulous, detailed records and monitors his or her playing results with scientific precision over a statistically adequate large sample size of hands played, the impact of making a correct adjustment may be functionally invisible. An impatient player may be tempted to abandon any experiment with the new, correct style of play rather than stick to it over months and even years of subtle, imperceptible, or yet-to-arrive benefits.

In the same way, we may lack the tenacity to restrict our play under ordinary circumstances to starting cards, with a reasonable likelihood of producing a favorable outcome against typical opposing hands. Our research may have told us that we're usually better off, when entering a pot from early position, if we limit ourselves to playing only pocket pairs, ace-king, or ace-queen, for example. But in actual practice, our craving for swift, short-term, hard-to-miss reinforcement of our decisions can cause us to lose patience during the protracted doldrums when we are dealt few if any of these pre-determined playable hands. We instinctively want action, and we want obvious and immediate positive results from our efforts. If nothing else, we have a psychological need for confirmation that we've made the correct decision, and that our investment in finding and implementing a new plan of action is paying dividends.

It takes a large measure of discipline and persistence to stick with our carefully-crafted plan when hour after hour crawls by with us sitting passively at the poker table, folding hand after hand, while action and excitement scream at top volume on all sides. As hours and even entire days drip boringly, snoringly by with precious little payoff apparent from all our patience, marginal starting hands such as queen-nine and jack-eight begin to look awfully tempting to us. And when we fold our cards, choosing not to bet and not to get involved, there is always the possibility that we may be missing the one pot we could have and should have won that would have spelled the difference between glory and shame for our entire tournament. That's right. I've said it before and I'll say it again in the next chapter, because it's such a fundamental point. There is risk in *not* getting involved and *not* betting, just as there is risk in taking action. If we muck our cards pre-flop and keep our chips stacked neatly in front of us and (we think) safely out of harm's way, we could actually be losing our opportunity to catch a miracle flop, bust two or three other players, double or triple our chip count, and greatly increase our chances of surviving all the way to the final table.

This is very much akin to the problem many people have in recognizing a broadly-dispersed and slow-acting environmental hazard such as mass extinction. If we are unable to see for ourselves dramatic evidence of either the problem or the gains from our corrective actions, we may well decide the trouble isn't worth our trouble. We don't get involved when we should, and we fail to avert an *avoidable* disaster. At the poker tournament, if we play too cautiously and don't take an occasional chance on a speculative hand, we run a real risk of missing our (well-disguised) golden opportunities to save ourselves from oblivion. We might never play the one or two monster pots that could have changed everything. So we just helplessly and passively watch as our stack dwindles and we eventually get blinded off. A poker player who only bets when she has the nuts or close to it will be too inactive to stay alive to the end. As the saying goes, in poker, "If you want to live you must be willing to die." Those cinch situations just don't come around frequently enough to allow us to survive, long-term. Just as in environmental choices, it can be equally as devastating an error *not* to invest in an uncertain situation—not to bet when the cards are questionable—as it is to waste our resources when the odds don't justify risking a gigantic proportion of our chip stack.

Or our impatience could finally overwhelm us and ruin us in a very different way. We might decide that we may as well have the fun of playing more hands instead of folding our way to oblivion, and play recklessly out of sheer pent-up frustration. We abandon our optimal starting hands and bet it up with almost any two cards, and we may even enjoy some big short-term wins from our new aggressiveness—additional validation of our decision to give up waiting in favor of wading in. We bet, and bet big, when the situation really doesn't warrant it. We bluff too often, chase speculative hands too many times, and invest far too many chips in pots where the odds are against us. This ultimately causes us to waste most or all of our chips because we lacked proper choices on when to bet and how much to invest.

With environmental problems, too, there can be a powerful urge to do something, anything, that seems directed at an apparent problem. If we take action in any form, then we feel intuitively that we have gotten involved and not just sat passively by, fiddling while our planet burns...or at least gradually grows dangerously warmer. Political leaders are particularly prone to succumb to the impulse to get actively engaged—and be *perceived* by the voters as getting actively engaged—to solve a matter of public concern. Psychologically, action itself is a potent reward. That is the advantage of immediate-gratification action over gradual, incremental tenacity: the showy, sudden reward instead of the shadowy, sloth-like deferred payout. It isn't mathematically sound or logically rigorous, but it definitely feels good enough to make us pay attention. And so the uncertainty, slowness, and subtlety of the right decision can lead us in the wrong direction, squandering resources in situations where we shouldn't have been involved if only we'd handled the uncertainty differently.

As in the environmental/scientific context, the poker-playing decision-maker has the power to reduce part of the uncertainty. She can do some research, study, gain experience, and carefully consider the probability and improbability of various events. By proactively investing considerable time, effort, and money into this collecting and processing of information, we can significantly promote the grade level of our educated guesses. For opponents in general, but especially with regard to certain specific individual challengers, we can narrow down the doubts quite a bit to such variables (among many others) as: (1) what range of starting hole cards they play from each position; (2) how much they usually bet when first entering the pot with specific hole cards; (3) whether they tend to slow-play monster hands; (4) whether they tend to check-raise after the flop; (5) how and to what extent they like to bluff; (6) how aggressively they play marginal hands such as one pair; (7) the degree to which they try to steal the blinds from late positions; (8) the circumstances under which they are willing to push all-in with various hands; (9) their tendency to go on tilt when upset; (10) how apt they are to lay down a decent hand if we bluff with a large bet; and (11) whether they can get away from a hand in which they've already invested a significant fraction of their stack.

We have the means to combine this type of study with the more formally mathematical variety, which is also important. That category would include the relative frequency with which each starting hole cards appear, the relative strength of each such cards, and the likelihood of any particular hand winning against a specific opposing hand. A player who combines solid grasp of the mathematical foundations with a skilled appreciation of the playing tendencies and tells of his or her opponents can greatly increase the probability of a successful outcome. But no amount—*no amount*—of preparation can definitively guarantee success over any span of time. Even in what we think is the long run, our long run through the fog may earn us nothing more valuable than stress fractures, foot blisters, worn-out shoes, and hemorrhoids. Thanks for the souvenirs.

In poker, we can never truly know the hole cards of our opponents until and unless they are exposed; we might have very strong suspicions, but that's all. We can never be sure what card will come off next from the dealer's deck; even if our opponent has only one "out" left among all the dozens of cards remaining in the deck, that card could be the sick one just waiting to hit the felt—and hit us right between the eyes. We can never have perfect knowledge what a specific opponent will do if we check-raise him all-in, even if we've seen him react the same way 30 times in a row. He could have recently read a book (imagine that!), or talked with someone, or just done a self-assessment, and now has a new game plan ready to spring on us. We can never sue the casinos if they've been grievously behind schedule in dealing us pocket aces, top set, an ace-high straight, a nut flush, or aces full of kings. No one ever signed a contract with us specifying just what "in the long run" means, and providing us with liquidated damages in the event of breach.

These and numerous other vital facets of the game are not just unknown—they are unknowable, now and forever. They are akin to some of the salient features of the mass extinction and climate change puzzles, perpetually beyond our reach like the grapes of Tantalus. Deal with it. That's poker. And that's life.

CHAPTER 10

Now *This* is a Gambling Problem

If you just picked up this book for the first time and skipped instantly to the final chapter to sneak a quick peek at a nice, concise, bottom-line Answer To Everything, I've got a friendly little poker game I'd like you to join. Anyone who expects a simple, easy answer to the biggest questions in Earth's history is a sucker. So if you're trying to cheat your way to saving the world, your heart might be in the right place, but your skull is full of mush. You're looking in the wrong book...or at least the wrong chapter of the right book. Snap out of it! Stop right now and go back to Chapter 1. There are no shortcuts to winning in this high stakes game—none that work, in any event.

If you've done your homework and read this book from the first page all the way through to this one, you should understand that there's no neat, tidy, insta-fix, happy-ending resolution to all the conflicts and questions I've set forth. You are missing the point completely if you've read this entire book and still persist in believing that today's Earth-shaking environmental dilemmas have an obvious single correct answer, and that we can tell the good guys from the bad guys by asking whether they accept the one and only true path to saving the planet. This isn't the type of book that would pretend there's such a facile fix available, one that can suddenly slice through our global Gordian knot of tangling issues. You should only be able to find that extreme level of oversimplification in a children's book, and not a very good one at that. Go ahead and flip through this volume one more time. Do you see any pop-up pages?

This is not a book about easy answers. It is a book about hard questions. More specifically, this is a book that searches for the wisest and most rational *approach* to the most difficult and important questions about our planet's wellbeing. I've spent a great deal of time exploring the many ways in which scientific and factual uncertainty permeates the vital issues of biodiversity loss and climate change. I've described in considerable detail the numerous instances in which doubt, incomplete information, and probabilities complicate our personal decision-making in some of our most crucial life choices. And I've tried to sift practical insights from paradigm-piercing work of some of the most creative geniuses ever to contemplate the nature of reality and our limited ability to pin it down.

With so many question marks, unknowns, doubts, gaps, and uncertainties forming the raw materials of this book, it should shock no one that the final product is not an exclamation point, but rather, another question mark. I'm going to end with this concluding chapter that does *not* pretend magically to blow away all of that fog, because past the fog we only find clouds. Instead, I will leave you with a way of asking the question that creates the best chance of making a reasonable, realistic, balanced, and objectively rational decision. No question can guarantee that the answer will be indisputably "correct," if in fact we can truly and definitively judge anything within our world's constraints to be correct. But we can *take steps* to get the odds in our favor and minimize the chance of making a foolish and devastating mistake. That may not be the ultimate, no-more-questions, capital-A Answer you were hoping for, but given the limitations placed on us by that annoying little thing known as reality, it isn't bad. As we say in the poker community, the most any of us can do is to get our chips into the middle of the table with the best of it. We gather all the information we can. We patiently wait for the right opportunity. We stay calm and unemotional. And when we bet, we do everything possible to have the odds as strongly on our side as the situation allows. The rest is out of our hands.

The Precautionary Principle from international environmental law is only marginally useful as a guide to making sound decisions. Beyond acknowledging the commonsensical point that we shouldn't insist on scientific certitude as a prerequisite to taking action—a necessary but not particularly stunning insight—the Precautionary Principle doesn't help us determine when we've got a situation that, in fact, calls for our intervention. Depending on the predispositions of people in power, it can fairly be interpreted as an open invitation to an aggressively proactive and perpetual "better safe than sorry" posture. Conversely, it could be taken as a nearly worthless statement of the obvious that doesn't actually mandate a response regardless of how strong the evidence becomes. For purposes of this book, I see the Precautionary Principle's primary value in its official recognition of the existence of uncertainty in environmental matters (and implicitly the very widespread incidence of such uncertainty), plus its rejection of doubt as a universal excuse for inaction.

That might not seem like much of a contribution, but in the cynical, self-interested, show-me world of environmental politics it actually provides a long-overdue whiff of reality. For hard-bitten, bottom-line types who only reluctantly believe even the evidence they see with their own eyes, it is a difficult first step. Is there truly room in the citadels of power politics for legislative action based on our planet's shaky, improvisational dance with uncertainties and probabilities? Can policy-makers peer beyond their obvious and immediate narrow concerns, and the similarly self-centered demands of their stakeholders, to see something in the foggy distance worth noticing?

It might help us in getting to "yes," or at least in getting to a skeptical "maybe," if we could imagine a trio of rather well-qualified consultants intervening. I've already introduced these modern-day "Three Wise Men" to you and said a lot of nice things about them. Now it's time to put them to work. Let's try to envision Pascal, Heisenberg, and Gödel alive right now, seeing what we see, knowing what we know, and facing the same information deficits and unknowns we face. We might as well make it as pleasant as we can for our distinguished guests, because they've traveled such a great distance to be with us today, so let's put them at a small table where they can enjoy a few snacks and beverages while playing a little poker. Okay, for Gödel's benefit we'll allow his wife to be there, too, so she can taste his food for him and reassure him that it's safe to eat. What might Blaise, Werner, and Kurt think about the dilemmas of biodiversity loss and climate change?

We can all agree that our three guests would never claim to be experts on climatology, ecology, biodiversity preservation, or environmental protection. But, like most of us (and yes, that includes me and you) that wouldn't stop them from forming and expressing an opinion on the day's issues. Famous people never hesitate to share their views on controversial matters, no matter how far removed they are from the fields in which they earned their fame. Being famous, it seems, gives a person access to a microphone with which to spread opinions no better rooted in fact or rationality than those of the typical person on the street. The difference is the microphone. So if an actor's or a comedian's or a rock star's opinion on Earth's fate is worth listening to, why should we refuse to consider the ideas of three of the greatest thinkers ever to draw breath? If lifelong politicians or professional heiresses can get people to pay attention when they talk about saving our planet, why not actually listen to a path-breaking, world-class, all-generations genius and take him seriously when he tries to join in?

At least, unlike the situation where movie stars, politicians, heiresses, and musicians pontificate on the environment, Gödel, Heisenberg, and Pascal actually worked on the highest possible level with some of the most exquisitely difficult concepts human beings ever contemplated. And their immortal contributions, at an absolute minimum, did in some way touch on aspects of doubt, incompleteness, lack of provability, uncertainty, and the limits of our ability to know. I admit that I'm imposing on our guests a bit by asking them to stretch beyond the precise, rigorous confines of their illustrious work. But at least I'm picking people who *know* how to stretch within the fields where such epistemological constraints profoundly affect reality, rather than the wildly irrelevant realms of celebrity and entertainment. When Gödel, Heisenberg, and Pascal stretched, they stretched our whole world right along with them. So let's hear what they might have to say about this high-stakes poker game where we are betting the Earth.

In Chapter 7, I introduced the concept of the World Wager and its accompanying tool, the world decision matrix. I proposed how we could use those analytical aids to help us take an evenhanded, rational look at two of our most pressing global environmental dilemmas. Both for the biodiversity hotspots and climate-change questions, I suggested a possible method of framing the wager and filling out the decision matrix, but I didn't try to connect all the dots to draw an answer. Now, with the assistance of Pascal, Gödel, and Heisenberg, I think we're prepared to revisit those daunting issues to see if we can tell in what direction they might point.

The world decision matrix arguably shows that, in the case of hotspots, on balance it is far more rational to invest in hotspots preservation than to stick to the status quo.[1] If we decide *not* to institute an effective hotspots protection program, the best outcome that might follow is that we save the fairly modest amounts of money and effort we could have spent on that program, and can divert those resources to other, potentially worthy, causes. However, the worst possible outcome from this decision is catastrophic—a mass extinction of huge numbers of valuable, even life-saving or epidemic-stopping plants and animals, entailing the irrevocable loss of innumerable vital sources of food, medicine, and ecosystem services.[2] Conversely, if we resolve to "bet" on hotspots protection and launch a workable conservation initiative, the best results would be unimaginably high—the rescue of vast living treasures that will preserve and greatly enhance human life and all life on Earth for millennia to come.[3] The worst-case result from a bet on the hotspots would be comparatively insignificant: the expenditure of considerable but not exorbitant quantities of time, toil, and tax dollars on a biodiversity program that wasn't really needed, and the concomitant loss of the option to divert those resources to (one would hope) more worthwhile causes.

Is this in fact a decent guesstimate of the conclusions Pascal, Heisenberg, and Gödel might draw from the available information? I think an impartial observer would conclude that the hotspots wager and decision matrix deal fairly and reasonably with the unknowns inherent in the mass extinction debate. The hotspots wager acknowledges the possibility that any or all of the unknowns might actually not be worthy of our intervention, if only we could somehow discover the truth. By giving equal attention to both the best-case and worst-case scenarios for each unknown, this approach accepts the full range of possibilities hidden behind the fog of uncertainty.

The real value of the hotspots wager and its accompanying decision matrix consists of the realistic and rational weighing of options and their attendant consequences, all predicated on best-case and worst-case answers to our unanswerable questions. I don't think it reflects a pro-conservation bias to posit far greater positive and negative outcomes under one set of possible answers, but much more attenuated benefits and detriments from another. It seems to be a matter of common sense to find the prevention or enabling of a global mass extinction, and the concomitant fate of hosts of extremely useful, life-saving, and/or economically valuable living resources to be of greater import than the saving or spending of a finite and rather modest (by governmental standards!) quantity of money and effort.

This is no different conceptually from a judgment that you're better off paying for insurance on your home, because if a fire happens, being able to rebuild is a higher-order consequence than saving or spending a much smaller amount on insurance premiums. If your house doesn't burn down in any given year, it turns out *in retrospect* that you didn't actually need the insurance for that year, but it would not be rational to conclude that the premiums you paid were wasted. You could not have known *in advance* that there would be no fire last year, considering that ruinous residential fires are known to occur with a rather low but not insignificant frequency every year. And what if you become disgusted with the regular annual expense, with no fire to make it all pay off, and decide to cancel your policy for next year? You are risking a huge amount (i.e., the loss of your house with no insurance money to replace it) to save a comparatively negligible sum of money (i.e., the cost of your insurance premiums). Pascal would be appalled. He would wonder whether you'd ever heard of his Wager at all.

This insurance analogy resonates with many people. That's why, in the decision matrix, I use the term "unused insurance" rather than "wasted money," referring to the fortuitous situation where unknowns were in reality not sufficiently threatening to warrant our expenditures to intervene.[4] That might turn out to be the case, but just as in the house fire example, we simply cannot know that in advance. We surely can't prove it. It's the type of information that is maddeningly missing from our incomplete little world, as Gödel would appreciate. Kurt would say that we shouldn't expect to be able to prove everything using the information and tools available to us, even within mathematically consistent systems. It is not a good reason to ignore the simmering crisis in the hotspots. It would be more akin to a reckless gamble than a wise investment to risk so much to win so little. It would be analogous to a poker player moving all-in against a taller chip stack in an attempt to steal the inconsequential blinds and tiny pot in the first level of the World Series of Poker Main Event. Why gamble everything to try to win next to nothing?

Of course, there are other ways of looking at the biodiversity issue, even if we agree to use the World Wager and decision matrix as our guide. I'm not attempting to claim that there is only one sane, or moral, or intelligent way of analyzing the situation. Depending on how we make our assumptions and what probabilities we assign to the various possible outcomes, we might reasonably conclude that intervention would be a bad decision. If our take on the facts and the unknowns leads us to assume that this notion of a modern mass extinction is much ado about almost nothing, and that the price tag of intervention could spiral out of control once we embarked on our ark-building campaign, we could wind up with an alternate ending for our story. I don't dispute that. I only suggest that it is crucial to make a full effort, choosing our parameters and our assumptions wisely and accurately. I'm just asking for maximum reliance on objectively verifiable facts and reasonably realistic presumptions, with minimum allowance for irrational prejudice or emotional conclusion-jumping. Fair enough?

Now let's consider the World Wager as applied to climate change. In Chapter 7, I selected three variables to juggle in the climate wager, and it took considerable effort to keep the number of unknowns that low. Based on the most extreme values each of those three unknowns could take, I then assessed the probable consequences of every permutation. The climate decision matrix summarizes those results. At this point, let us look at that matrix again, still with the imaginary help of Pascal, Heisenberg, and Gödel looking over our shoulder during a break from their poker game. With their advice, let's see how the climate matrix incorporates and even harnesses the uncertainty inherent in the climate change conundrum.

Our contemporary re-make of the "Three Wise Men" might argue and disagree among themselves on the meaning and significance of many facts and assumptions underlying the climate change puzzle. Each of them is in some sense a fish out of water when it comes to global climate modification. In that regard they have plenty of company among the legions eager to expound their pet views on this controversy. Gödel might dismiss an opinion from one of his companions, scolding him, "Werner, you have no qualifications to tell me which way the wind blows. I'll look you up when I want to know how to find an electron!" Heisenberg could counter, "Kurt, you've spent too much time with Einstein. What do you know about the real world? Go back to your numbers and see if you can get them to add up!" Pascal could vent his frustration at both of them: "Gentlemen, you may have written the book on mathematics or quantum mechanics, but you should have the humility to admit that your work has its limitations...and this question is everlastingly beyond the pale for both of you." Gödel might then fire back, "Stop praying long enough to open your eyes, Blaise! On climate change, people are listening to opinions from famous celebrities whose qualifications begin and end with being the Prince of Wales or the heiress to the Hilton fortune. Don't you think I deserve to be heard too?"

Let's skip the ensuing Jerry Springer Show free-for-all and examine what our three friends might say once they were finished excommunicating one another from the Church of the Expert Opinion. They could focus on some valid, if sometimes surprising and inconvenient, points we can extract from the climate decision matrix. I emphasize again that this is absolutely *not* the only way to look at the matter, nor is it necessarily even the best, most rational, most intelligent, or wisest. This book has been all about preparing every reader to make that judgment independently, equipped with some useful tools and a reasonably adequate collection of relevant facts. Here then, is *only one* credible set of conclusions made possible by a systematic consideration of the decision matrix.

It doesn't take long to realize that, in comparison to the hotspots matrix, there are more positive outcomes related to non-intervention in the climate matrix. Six out of the eight possible permutations of the unknowns—if we decide on non-intervention in climate-change—yield "lucky wager" results, as opposed to just four in the hotspots situation. This difference stems from the uncertainty regarding whether it is even possible for human beings to make the right kind and amount of change in Earth's climate, irrespective of how important it is or how hard we try. This doubt as to the possibility of successful intervention itself is a key point of divergence relative to the hotspots matrix. With biodiversity preservation, there is no legitimate scientific or practical question that we have the capability of getting the job done if we decide, just this once, to try.

We know, more or less, what it would take to conserve the hotspots—the top enclaves of endangered life on Earth—as viable habitats for the many species huddled within them. Because we happen to be fortunate enough to have the overwhelming majority of all threatened species conveniently concentrated together in these cartons of eggs, so to speak, we have a fairly defined task at hand if we decide to protect the eggs Nature has put in one basket. We understand what this would require, from strict controls on deforestation and other forms of habitat alteration, to adequate policing to prevent poaching and introduction of invasive species. These measures have worked on many other key habitats, large and small, all over the world; and there's no reason to doubt they would work just as well with the hotspots habitats. There's really no great mystery about it, just apathy and lethargy. We know it can be done, and we know how to do it. We've just never cared to try.

In contrast, the realistic prospect of failed intervention in the climate change context interjects a formidable additional dose of uncertainty into the decisional equation. If it is objectively impossible for us to change the course of our planet's climate, then no matter how much we might need to or want to, we'd be better off focusing our attention and resources on those things we can actually influence. It would be a colossal waste to expend enormous amounts of money and effort to build a superhighway to a mirage in the desert; instead, we could devote our resources to bringing in water and finding ways to keep the desert from expanding.

This question of impossibility has at least two components, and both of them are substantial. There is the element in which it might be *literally* beyond the technological, scientific, and engineering capabilities of the entire human race to make a purposeful and directional shift in Earth's climate...even if we are all united and fully committed to this. That is, even if every nation of the world stands as one and makes a good-faith collaborative push in the greatest example of global cooperation ever, it may not work. Nothing like this has even been done, and nothing like this has ever been tried. For a meta-system as complex as this planet's entire climate, there is no such thing as a clean, simple, one-cause/one-effect fix. Multiple interlaced factors combine to nudge temperature, cloud cover, wind speed, ocean currents, and storm systems in various directions over time, and there are built-in mechanisms that can serve to counteract perturbations and alterations. Unlike the classic scientific experiment in which we mostly hold all conditions constant except for one variable we choose to adjust, our planet has the irritating trait of never holding its myriad conditions constant. Earth just won't hold still for us. There is no way to ensure that our extraordinary efforts to reduce human-generated greenhouse gas *emissions* would produce any particular reductions in the *amount* or *concentration* of such gases in the atmosphere, or that such reductions would in fact translate into any certain diminution in global mean *temperature.*

But there is also the possibility that we could, *in theory,* effectively push back on climate change if substantially all of the world's nations worked together in concert ("See it now! The Human Race, live and in concert on their sold-out 'Cool It' worldwide tour!"), and nevertheless we would never be able to cooperate *in practice* to the necessary degree. Given the deep and enduring divisions based on religion, ideology, political system, race, ethnicity, historical conflict, economic system, level of economic development, competition, and a litany of other splintering factors, it may be functionally impossible to achieve the necessary international cooperation. Would enough of the major powers and sources of greenhouse gas emissions truly put aside the impulse to pursue competitive economic advantage in this increasingly congested world market? Would developing nations and developed nations be able to collaborate on a cap-and-trade regime that affects the two sides in very different, but extensive ways? Would the world find the will to "stand as one" and overcome the free-rider problem or the tragedy of the commons when there is so much that divides us?

This human factor is at least as troublesome as a source of uncertainty as any aspect of physics or mathematics. For highly focused and specialized geniuses like Gödel or Heisenberg, who had somewhat limited ability to deal easily with other people on a person-to-person level, human unpredictability may be more of a wild card than any scientific variable. Indeed, how can anyone confidently predict an unprecedented planet-wide communal barn raising to chill a supposedly overheating planet? From the League of Nations to the United Nations and beyond, there are few if any inspiring examples of unity or unanimity among humanity to give us realistic hope that it would happen this time, for the first time. Unless you count fictional examples from pop culture such as the film *Independence Day*, the cupboard is conspicuously bare of instances of give-till-it-hurts line-crossing global cooperation.

Even within the single nation known as the United States of America, there are powerful stressors, tensions, competing interests, conflicts, differing conditions, and internecine rivalries. A carbon-focused program to throttle back on our greenhouse gas emissions could disproportionately burden states that are major producers and/or consumers of coal (such as West Virginia, Ohio, Pennsylvania, and others). The program could put much more hardship on states where a lot of jobs depend either on mining and transporting coal, or on heavy industries that rely on coal to fuel their processes. And the states where coal provides most of the energy would shoulder more than an equal share of the costs in the form of higher prices for energy, in homes, schools, businesses, and factories. These and other splintering impulses are major impediments to formulating a workable global warming strategy even within the United States of America. When we try to take this show on the road and expand it to the worldwide level—the Disunited States of Earth—those fracturing forces become even more irresistible.

Nations that lean heavily on nuclear power for their energy and/or produce very little coal are in a very different position from those in the opposite situation where an impactful cap-and-trade campaign could inflict massive harm in the form of increased unemployment, lower productivity, worsened balance of trade, and shockingly higher energy costs. Petroleum-exporting nations would also be burdened to a much greater extent than service-industry or technology-heavy countries that are relatively uninvolved with fossil fuel production or consumption. Wealthier nations with stable or declining population and well-developed, modern infrastructure already in place are also in a decidedly distinct posture compared with rapidly growing, largely poor, still-emerging nations with economies just getting off the ground. It's a noble ideal worth aspiring to, that all nations would put aside their narrow self-interest and competitive instincts in deference to the common good of the planet. To say it's doubtful that it will actually happen—at least at a proactively early stage before the crisis becomes blatantly obvious and functionally irreversible—is a bit of an understatement.

The climate change wager also differs notably from the hotspots version in the magnitude of both the "lucky wager, money saved" and the "unused insurance" outcomes. Both of these results appear to be far larger in the climate change context, and their significance is therefore greatly intensified. This follows from the fact that it would be much costlier—in multiple different ways—to neutralize any major climate change threat than to halt the loss of the biodiversity hotspots. Even assuming that it is indeed possible for us to exert adequate targeted influence on our planet's climate (which, as we have seen, is by no means certain) it would by all accounts be extraordinarily expensive. It would probably require massive international transformations of lifestyle, industry, energy production and use, economic activity, transportation, land use, and an array of additional fundamental features of human life. And these transformations would have to be permanent, not transitory. Now it may well turn out that circumstances show it to be worth all that hardship to turn the tide on climate change, but nonetheless it would be an expense—or an investment—of absolutely unparalleled scope and magnitude.[5] This means that if we make all those changes needlessly, or ineffectively, we have wasted unimaginable quantities of resources.

Pascal and his friends might note that action of these breathtaking proportions has few if any obvious parallels among our everyday life choices. Whereas intervention to preserve the biodiversity hotspots would be proportionally no costlier or more disruptive than any number of other rather routine governmental projects, the same cannot be said with a straight poker face about dialing down the global thermostat. The profound, multi-faceted, fundamental transformations entailed in climate intervention are well beyond the level of expense and life-changing consequences associated with any ordinary form of life, health, or home insurance. If we insist on adhering to the insurance analogy, it would perhaps be most similar to an attempt to buy life insurance when we already have clear symptoms of a nearly-incurable and often-terminal illness at an advanced stage.

When you try to buy life insurance with one foot in the grave, the price is enormously higher...if insurance is even available at all. This is because life insurance is designed for a routine precautionary situation, and not an actual emergency already in progress, with all the urgency and desperation that implies. If you could buy life insurance at any price, it would have to be incredibly costly to reflect the gravity and imminence of the crisis. You would probably have to sell your car, mortgage your house, and make other major sacrifices to be able to afford the astronomical premiums. Pascal would observe that a choice demanding this much of a person is different in kind from the one in his famous Wager. Instead of being fairly easy, painless, routine, affordable (or even money-saving) and accompanied by tangible positive side-effects, the interventionist option under the climate issue is extremely difficult and expensive on several levels, requiring multiple core-shifting re-workings of the pillars of modern society.

It's easy to get lost in the ocean of possibilities and what-ifs as we try to place some realistic boundaries on the costs and benefits of climate change intervention, as well as the magnitude and credibility of the threat itself. The complexity of both the problem and the possible countermeasures is almost beyond the reach of our comprehension. This conundrum, of course, was made much worse by the late-2009 revelations regarding evidence destruction, suppression, distortion, manipulation, and exclusion at the Climatic Research Unit.[6] By undermining our confidence in the integrity of some of the key underlying data and trends, this "Climategate" scandal has injected new uncertainty into an already-uncertain equation. At the risk of going too far along a speculative detour, however, allow me to mention one additional possible consequence of a full-scale remediation effort. Strangely, this eventuality can plausibly be viewed as either a very positive or extremely negative development, depending on one's perspective; so it becomes yet another element of uncertainty to factor into our calculations. Whether it constitutes an added reason to err on the side of intervention or an extra argument counseling restraint is a matter that can only be resolved by the people making the decision. Heisenberg, move over—*this* is uncertainty!

I'm referring to the prospect of some form of overarching world government emerging as part of the global campaign to reverse climate change. This has been discussed as a conceivable step on the road to the unprecedented level of multi-national cooperation that would be required to implement meaningful climate countermeasures. If we might only have a realistic chance at halting harmful climate change with all or nearly all the world's nations pulling together, and if the degree of commitment would be much higher than that needed for any prior cooperative effort, the crisis could be the catalyst for formation of a government with truly global reach and power. If the amount and type of cooperation is beyond the limits feasible within the framework of the United Nations, the European Union, or any other gathering of peoples, and if the magnitude of the threat is perceived as warranting it, a world government might be seen as the only option. Nations are irrelevant if humanity is destroyed, so maybe this is the way of the future—a single unified world, with many states within it, coexisting and collaborating in a spirit of mutual support. How else can we overcome the free rider problems and the various reasons why key players want to sit this one out?

You can see why a massive, worldwide mobilization to combat climate change at least arguably could be the precipitating cause of a move to world government. National sovereignty and the independence of individual nation-states are genuine obstacles that impede any coordinated global enterprise, and all the more so when the project in question demands extraordinary sacrifices, fundamental reconfigurations of society, and profound shifts in the way the world works and lives. Particularly where the campaign may disproportionately advantage or disadvantage certain nations or territories to a significant extent, it may be impossible to get all the necessary contributors on board without the coercive power of a supreme, legally legitimate, substantively potent superior world government to enforce universal compliance. Where individualized self-interest is the centripetal force tending to pull nations apart, what except a United States of Earth can pull them together? An umbrella government can compel its component units (formerly sovereign nations) to make whatever specific set of sacrifices and adjustments it deems to be in the best interests of the world as a whole...subordinating the needs of the few or the one to the needs of the collective many.

Clearly, the abolition of national sovereignty and the establishment of a powerful, legally effective world government is about the most momentous and profoundly transformative human event imaginable, second only perhaps to the prospect of the Chicago Cubs winning a World Series.[7] Whether it would be a blessing or a curse is another matter entirely, and that issue is bull's-eye-middle in the eye of the beholder. Some would view the advent of world government as the dawn of a bright new post-national era, free from the divisive and destructive wars, rivalries, sanctions, tariffs, persecutions, and oppressions that have so badly marred and bloodied all of human history since the rise of the nation-state. Others would see it as a death warrant for freedom, individuality, religious liberty, cultural traditions, economic experimentation, and varied approaches to government, with no real efficacious counterforce to the tendency of an all-powerful central government to gravitate toward totalitarianism. Irrespective of whether the outcome would be closer to one extreme or the other, the possibility of a worldwide climate change mobilization facilitating the rise of a world government is one more very big wild card any decision-maker would need to factor into the equation. One way or another, this issue greatly intensifies the already-stupendous size of the wager we would be placing by betting in favor of climate change intervention.

The meaning, then, of a lucky wager or unused insurance outcome in the climate context is arguably different in kind, and not just different in degree, from superficially equivalent results in the biodiversity context. It's akin to the difference between making a small raise at the poker table and pushing all-in. Instead of risking a fairly small, limited quantity of resources to win a pot, we are gambling a huge amount—putting our tournament life on the line, as they say—and in effect placing in jeopardy the viability of our current economic, political, and social system. There are, of course, times when an all-in wager really is the best alternative available under the circumstances. But it matters, in terms of how carefully we need to weigh our options, whether we are betting a little or an enormous stack of chips. Again, if and when all the evidence on all sides is fully, fairly, and objectively sorted out and cleared of partisan distortion, the situation may legitimately call for an all-out campaign to halt global climate change, with nothing held back, because the consequences of failure are so unthinkable. But this extreme commitment requires meticulous analysis of all probabilities, risks, and rewards. It is not in the same league as one of those commonplace, routine decisions we make reflexively as a necessary cost of doing business in the real world. Therefore, it is a matter of tremendous weight that we make this decision with precision and selectivity, only when needed and only when it will actually make a major difference for the better. A wrong choice here is much harder to write off as just one more of the normal expenses we regularly incur and take in stride within our "better safe than sorry" standard operating procedure.

Pascal would know that probability theory, in poker and in life choices, sometimes calls for inaction: not betting, not buying insurance, not investing, not getting involved. The cliché "A penny saved is a penny earned" does not become invalid when we change it to "100 trillion pennies saved is 100 trillion pennies earned." At the poker table, sometimes the best bet is no bet, and the best way to play a hand is to toss the cards directly into the muck without risking a single chip. Other than blinds and antes, we are never forced to bet or to play any particular hand. Once we choose to bet and see what happens with our cards, many possibilities suddenly open up to us... some of them very good, and some of them very bad. You can never lose all of your chips, though, if you just opt out of any given pot. You might miss out on a big win, but at least you know you won't be sent to the rail with a tournament-ending loss. Pascal would urge us to consider the upside *and* downside potential of each alternative, both action and inaction, logically and carefully before we decide whether to play, pay, and pray.

Gödel might point out that, within our system of incomplete information, it is also immensely important that we do exercise the intervention option when called for. He would acknowledge that our imperfect knowledge prevents us from securing a complete understanding of all the relevant and pertinent facts, but that there are some permutations of unknowns that would make it disastrous to guess wrong in favor of non-intervention. Where the (unknowable) truth is that there will be massive devastation brought on if current climate trends are not reversed, *and* where the (unknowable) truth is that we do have the capability, in theory and in practice, to reverse those trends adequately, it would be catastrophic to remain on the sidelines. The combination of both of these circumstances, joined together, is at the heart of the four situations in which an all-in bet is correct, and it's a huge mistake to hold back and wait to play another hand. This could apply even if the worst-case catastrophe is highly improbable but does have a non-zero probability. To put it bluntly, even if the late-2009 "Climategate" scandal significantly diminishes our estimation of the threat, taking precautionary action still *might* be the best bet under some circumstances.[8] We'd be betting the Earth one way or the other, whether we launch a full array of climate counter-measures or decline to act. The difference is in the results of that betting decision. We either win a gigantic pot (by staving off the carnage that would have followed unchecked climate change) or miss the one hand in which we could have turned our entire tournament around (by failing to take action when conditions were right).

Both Heisenberg and Gödel would understand that, in our modern, complex, litigious world, we can unleash chaos by *refusing* to choose an active approach to a monumental challenge like climate change. In Chapter 2, I mentioned that the 2007 Supreme Court decision in *Massachusetts v. EPA* had launched an unpredictable game of hot potato to decide the United States' response to global warming. Because the elected leadership of the nation had chosen not to be proactive on climate change, a collection of interested parties were able to go to court. Once inside the very different parallel universe of the judicial branch, they used their "home court advantage" to the maximum, in a veritable full court press. They won a judge-made result that forced a reluctant administrative regulatory agency to tackle climate change using a statute, the federal Clean Air Act, that was never designed for that type of issue.

And now, cut loose from the pressures of elections and the political accountability of the legislative branch, the climate change juggernaut could resist even the attempts of the executive branch (through its Environmental Protection Agency) to control it. The EPA's leaders, guided by the president, may want to use a "tailoring rule" or other administrative authority to decide on a measured, pragmatic program to regulate greenhouse gasses and the climate change puzzle...but the courthouse door is still wide open, and the lock is broken. Using the Clean Air Act's citizen's suit provisions, environmentalist groups or individuals can sue and try to get the courts to compel EPA to apply every statutory provision to greenhouse gasses, fully and without modification. Unelected, life-tenured judges may well agree with the litigants that the Clean Air Act means exactly what it says, and that EPA *must* treat sources of greenhouse gasses the same as all other sources of "regulated pollutants," even if the standard 250 tons per year threshold is obviously unrealistic.

When people whose duty it is to make decisions fail to make them, or choose options without adequate care and fairness, they send away for just this type of Pandora's Box. As the saying goes, "Choose or lose." Or maybe that should be "Choose, or lose the power to choose." The greenhouse gas issue that the Supreme Court let loose in *Massachusetts v. EPA* proves that the wrong approach to uncertainty in decision-making can lead to loss of decision-making control. Now, we have the courts and private citizens wielding real power on climate change, not just in addition to but even in opposition to the legislative and executive branches. All three branches of government are involved, and riding on the judges' robes are armies of eager litigants from the private citizenry. That's an enormous number of cooks trying to influence what kind of broth we'll serve on this hot issue.

The World Wager and decision matrix can help put the lid back on this climate-change Pandora's Box. The best way for rational decision-making to rise to the top in a multi-party scramble like this is for the most well-established, enduring, and officially empowered entities to exercise leadership. If the president and top congressional leaders take the initiative and proactively devote serious attention to messes like climate change, courts will be less likely to perceive any abdication of responsibility, and thus will be less willing to hand momentous victories to citizen's suits. These formal repositories of official authority are best suited for what the decision matrix requires: careful, thorough, comprehensive, expert-level analysis and consideration of all the factors, all the potential consequences, and all the probabilities.

Decision-making by litigation, in contrast, is almost inevitably haphazard, piecemeal, slanted toward certain pet issues, emotionally driven, and disorganized. An executive- and legislative-branch partnership that takes the climate change decision matrix seriously, and uses it honestly, should not find the judicial branch picking its pockets for the keys to the car and then tossing the keys to the crowd in the street. It's mainly when courts perceive a vacuum of inaction, an emptiness of thought, and an absence of sound decision-making by the political branches that they will intrude and let private litigants slip into the vacant driver's seat. Once they do, it's anybody's guess where we'll go. It's a double bind. Poor decision-making by those who are paid to choose wisely invites all of us amateurs to see if we can do any better. Three guesses on whether most judges and plaintiffs have ever heard of decision theory, let alone know how to apply it. So when no one is using anything resembling the World Wager, our typical mode of choice analysis seems almost designed to produce bad results, either in the form of needless meddling (damned if we do) or misguided inaction (damned if we don't).

Gödel would recognize strong parallels between this issue—the twin dangers of betting when we shouldn't and not betting when we should—and the work of two of his contemporaries dealing with statistics and decision-making errors. In fact, as our group debates the issue around their poker table, Kurt might want to call these friends on the phone, and "use a lifeline" so to speak. Well why not do it right now, as long as we're imagining all this intellectual firepower at our disposal? Let's allow Gödel to phone his friends and see if they have anything to add to this hot topic.

Around the same time Gödel introduced his incompleteness theorems, Jerzy Neyman and Egon Pearson unveiled their ideas on two fundamental errors inherent in any use of numerical evidence to decide whether a hypothesis is true. If Pascal were able to see the insights of Neyman and Pearson, he too would appreciate how they aid in our use of probabilities to weigh options among competing alternatives, and he might find a way of incorporating them into his legendary decision-making toolbox. So let's take a quick peek at what Neyman and Pearson had to say, and why it matters to the subject we have under consideration in this book.

The core idea is both simple and elegant. There are certain basic ways we can make a mistake when trying to assess the available statistical evidence and whether it supports a judgment favoring the proposition in question. That is, in a scientific experiment or any other form of methodical evaluation of options, there are major categories of how things might go wrong and lead us to an incorrect conclusion. The important point is to be fully aware of this and to stay focused on our core fundamentals, with logic and discipline. If we maintain control over our emotions and biases, and keep our mind on making rational choices on what risks we're willing to take, we can stay on track through all the shocks and rocks awaiting us in the fog ahead. When we *know* and *plan for* the errors we might commit and what they mean to our goals, we are less vulnerable to falling into that most terrible of poker player traps: Going on tilt!

During any poker game, there can be bad beats, long periods with unplayable cards, unpleasant surprises, difficult conditions, and all types of negative factors. I can assure you, as bad as these things can be, a player will make them immeasurably worse by going on tilt. Tilt is a condition known to everyone who has ever played poker. It's a case of emotional hijacking. It happens when we let our feelings overwhelm our thoughts. We go on tilt when we allow ourselves to become upset, angry, vengeful, despondent, impatient, distracted, or desperate to the extent that we lose control of our game plan. Instead of maintaining our carefully-assembled set of skills and professional attitudes to steer calmly through our difficult times, we abandon reason and react viscerally. Enormous stretches of fundamentally solid work and sound judgment, together with all the resources we've systematically amassed during all that meticulous concentration, can be undone in an instant of wild rage or outrage. When we're on tilt we forget everything we consciously know about what we should do, and substitute an adrenaline-fueled impulse-driven reflex...often with disastrous results. On tilt, we risk much more for the sake of much less than we normally would when our head—and not our gut—is driving the car.

To avoid the deadly tilt minefield, we have to anticipate what can go wrong and build in countermeasures to deal proactively with those errors and bad beats. This requires self-control and determination, plus the foresight to see potential trouble spots before we fall into them. As we go about betting the Earth, the worst thing we could do would be to lash out in an emotionally-unhinged erratic reflex. In this momentous enterprise, that means we need to come to grips with the main types of errors we can expect to encounter, so we can plan for them and shift the odds in favor of the results we most want to achieve. Think of the work of Neyman and Pearson, then, and my variation on their ideas, as a prescription for tilt control medication that can keep us from blowing up—and blowing it—as we gamble with the future of this planet.

Neyman and Pearson brilliantly distilled all the countless ways in which people can make mistakes (and believe me, because I've tried them all—there are a lot) into just two basic varieties. Their first category of blunders amounts to this: finding that the proposition being tested is true when in actuality it is false. This is known by several names, including Type I error, α error, error of the first kind, or false positive. In scientific or statistical language, the Type I error results in our rejection of the null hypothesis when in reality we should have accepted it. Putting it that way makes it sound a lot more complicated and confusing than it really is, but that's what happens to a guy who works too long as a law professor. You forget how to use plain English. Okay, let's back up and let me try that one again.

The null hypothesis is the default option or presumption that there is *no* actual effect, *no* causal connection, or *no* genuine phenomenon present in the subject matter we're examining. It's an assumption that there's nothing going on worth paying attention to in whatever we happen to be studying. Hundreds of my students have made this assumption about whatever I've tried to teach them, and I'll bet they didn't even realize they were making a null hypothesis as they drifted off to sleep in my classroom. The null hypothesis remains in effect until and unless we gather sufficient credible evidence to overcome that assumption. It's similar to the presumption of innocence in a criminal trial, where a defendant is presumed to be not guilty (that is, assumed to have done nothing wrong, nothing deserving of criminal punishment) until and unless the prosecutor can submit enough legal and competent evidence to prove him guilty.

So, in other words, when a Type I error induces us to reject, incorrectly, the null hypothesis, this means that it makes us *think* we've found a statistically significant effect when in fact there was *no* such effect. With Type I error, we believe we are seeing something meaningful when in truth we're only staring at a mirage. So instead of wisely and correctly walking away and concluding there's nothing worth getting excited about (that is, that the null hypothesis should be accepted), we invest time, money, and effort futilely and wastefully in further research and action to pursue this illusion. Type I error trips us up by getting us to mistakenly reject the null hypothesis and needlessly get all excited about the situation...a classic case of making a mountain out of a mole hill, or much ado about nothing.

This error of the first kind is a form of undue gullibility or susceptibility to false alarms, where we are inappropriately overeager to swallow weak or nonexistent evidence and jump into an exciting but fallacious conclusion with both feet. Some familiar daily-life examples of false positives include tests that indicate a woman is pregnant when in fact she isn't, or mammograms that suggest she has breast cancer when she doesn't, or prostate cancer screening procedures that point to the existence of cancer when a man is actually cancer-free. For anyone who has ever lived through any of these unfortunate situations, it's easy to see how Type I errors got the nickname "false alarm." The flawed information often touches off a firestorm of frenetic, costly, upsetting, and even panicked activity, all aimed at dealing with a non-existent situation.

The second category of error neatly complements the first. It consists of our acceptance of the null hypothesis when we actually should have rejected it—that is, we decide that the proposition in question is false when in fact it is true. This is called Type II error, β error, error of the second kind, or false negative. With Type II error, we disregard or devalue key evidence, miss it entirely, or misinterpret it, to such a degree that we can't see the reality of a threat, the soundness of a postulated cause-and-effect relationship, or the validity of a new theory. Type II error leads us to overlook statistically meaningful and genuine effects. We decide that we have been barking up the wrong tree when there's really a cat (maybe Schrödinger's cat?) hiding up there among the leaves and branches after all.

A false negative can push us away from the study and confrontation of important, even vital, issues under the mistaken belief that there's nothing there worthy of our attention. This error of the second kind is an example of excessive skepticism or carelessness that ends up in missed opportunities and unanswered threats, costing us the identification of real and significant phenomena. By inducing us to remain blind to genuine issues, Type II error can shove us out of bounds from the playing field and onto the sidelines where we then stand passively by while major events unfold with no involvement from us. We could illustrate with the same examples as before, but with the substitution of the opposite incorrect conclusion. That is, false negatives can come in the form of a test that does not indicate pregnancy when the woman is in fact pregnant, or a medical evaluation that fails to identify the presence of prostate cancer or breast cancer in a patient where the cancer is actually already there, extending its deadly tentacles.

Is either type of error generally worse than the other? It depends on the nature of the question under consideration and whether, in that context, we do more harm by acting needlessly or by refraining from action when it's called for. The examples I chose happen to tend to have more serious consequences in the face of Type II error than Type I. In these situations, a Type II error might cause a woman to take medications, smoke, drink, or engage in other physical activities harmful to the fetus she doesn't know she's carrying. Likewise, a man relying on his incorrect good-news prostate test results, or a woman misled by her clean mammogram, could easily decide there's no need to go to a physician until the deadly disease spreads so far that symptoms of cancer finally show up and become too severe to ignore. That delay in treatment brought about by Type II error could allow the cancer to metastasize unchecked to the extent that it costs the patient his or her life.

On the other hand, Type I errors can be extremely harmful as well. In circumstances where a false positive generates a course of action (or reaction) that is itself hazardous, very expensive, difficult, and laden with negative side effects, pulling the handle on a false alarm can be a very serious and even fatal error. For example, a patient erroneously diagnosed as stricken with cancer might undergo dangerous, pricey, and painful surgery, perhaps coupled with chemotherapy, radiation, or other treatments that also are debilitating, costly, and decidedly unpleasant. It's trite but true that the treatment can be worse than the disease, particularly where there is no disease to begin with.

Within the scientific community, Type I error is generally considered a more serious problem in interpreting the meaning of experimental results than Type II (although clearly no rational person wants to commit either blunder). Perhaps from an abundance of caution and professional restraint in declaring significant findings, scientific researchers prefer to err on the side of Type II. Again, there are parallels to a criminal trial, in which the defendant benefits both from a presumption of innocence and a requirement that the prosecution prove its case beyond a reasonable doubt. The legal principle—that it is better for many guilty defendants to go free than for one innocent person to be convicted—is consistent with the scientific aversion to Type I errors. A prosecutor's high burden of proof is analogous to the scientist's guideline: It presumes no significant result *unless* the evidence indicates the probability of it happening through random chance is less than one percent. Both presumptions, and the strictures they bring along with them, are intended to reduce the occurrence of Type I errors.

The dilemma, of course, resides in the tension between Type I and Type II errors. As we take steps to avoid seeing things that aren't there, we increase the chance that we will miss seeing things that are. For example, we might, in our war against Type I errors, adopt more restrictive thresholds, such as a strong presumption against reaching action-inducing conclusions. This would usually involve some version of a stringent burden of proof—a strong presumption in favor of innocence (i.e., in favor of the null hypothesis)—and rigorous requirements for evidence to be deemed acceptable. This strict filtering mechanism should succeed in reducing the incidence of Type I errors, but it could (and very likely would) also provoke an upsurge in Type II errors, with those false negatives causing us to overlook some genuine concerns. We would sacrifice our ability to spot some real positives in our quest to eliminate false positives.

Conversely, if we prefer to reduce the probability of false negatives (i.e., we want fewer Type II errors), we could impose more sensitive threshold filters and presumptions. A lower burden of proof, weaker presumptions against taking action, and more relaxed standards for evidence would help. Sensitive tests let more things pass through the barriers around us so we can pick them up and do something about them, where more restrictive screens might block them from ever showing up on our radar screen. This emphasis on sensitivity rather than restrictiveness will enable us to capture more actual positives, but at the cost of also finding more false positives (i.e., more Type I errors).

Does this sound like a grandiose case of "damned if you do, damned if you don't?" Or being "stuck between a rock and a hard place?" Well, that's because it pretty much is. Scientists just prefer to phrase it differently, because, well, it sounds more scientific when you polish it and dress it up a bit. But at the heart, it's a fact of life we're familiar with because we live in the real world.

In the real world, both types of errors will happen, in part because of the uncertainty and incompleteness we've been discussing in this book. With incomplete, imprecise, and imperfect information to work with, it is unrealistic to expect correct decisions across the board. We can't always identify every meaningful phenomenon wherever and whenever it exists and yet never mistakenly see cause for concern where there is none. That is, we can't be that vigilant yet still avoid all Type I errors. Now turn it around. If we are absolutely determined never to be duped into taking action needlessly—never fall for a scary illusion—we have to get used to letting some genuine threats slip past our defenses as well. Put differently, we can't be that measured and conservative in our responses, yet still dodge all Type II errors. Unavoidably, there is to some degree an inverse relationship between the two varieties of miscues: As our Type I error rate declines, there will be some rise in Type II errors, while an increase in Type I errors will be accompanied by a decline in the Type II rate. Any determination to dial down the prevalence of one type of error will come bundled with presumptions, evidentiary rules, burden of proof requirements, and other decision-shaping conventions that also generate a surge in the other, "lesser of two evils" brand of error. So what does all this mean when we are betting the Earth and don't know what bet to place?

Our three mathematically-oriented friends, Heisenberg, Gödel and Pascal, would be able to take the easy next step from the work of Neyman and Pearson and apply it to today's global environmental challenges, by way of gambling analogies. As they sit around our imaginary poker table, Heisenberg might lecture to and argue with Pascal and Gödel over how we could modify the traditional Type I/Type II model to suit our purposes. And although we could defend a decision to stay with the standard and familiar Type I/Type II nomenclature, our experts would probably favor adoption of distinct terminology, for the sake of rigor and clarity of meaning. Plus, it's always fun to come up with a new name for something you've invented. So after the phone call to Neyman and Pearson concludes, Pascal and Gödel might talk it over with Heisenberg, work through their differences, and suggest something along the following lines.

Within the context of gambling decisions, including the peculiar variety of gambling I choose to call "betting the Earth," there are two chief categories of decisional mistakes. They are similar to, but not identical to, the Type I and Type II miscues, and they have particular relevance to resolution of our ultimate game of high stakes poker. So let's coin a couple of new terms here and see if they can help us make these crucial decisions. We can call these Type R errors and Type P errors. There! That was easy.

Let's define Type R (R for restless) errors as those incorrect decisions to bet, or to bet too much, when conditions actually favor little if any involvement. Type R errors are the result of our hyperactive proclivity to take action when faced with an apparent challenge, avoid boredom, seek out positive change, and generally participate rather than merely observe. Conversely, we can define Type P (P for passive) errors as those incorrect decisions not to bet, or not to bet enough, when conditions actually favor vigorous involvement. Type P errors stem from our tendency to be overly conservative, passive, reluctant to take action, and wedded to the status quo. Type P errors are related to, but not the same as, Type II errors in that both reflect a penchant for conservatism in making weighty judgments. That is, a cautious and tradition-honoring preference to err on the side of being calm and level-headed, and, when in doubt, not making a dramatic break from the current practice or principle. Likewise, Type R errors share in common with Type I errors a predilection and impulse for being proactive, creative, involved, results-oriented, and energetically engaged in events.

From this explanation, you can see a definite, positive way of looking at the mindset or attitude that tends to produce either family of errors. With the best of intentions, and wrapped up in very defensible outlooks on risk, people can fall into either Type R errors ("sins" of commission) or Type P errors ("sins" of omission). There is nothing innately evil, ignorant, mean spirited, stupid, disreputable, or dishonest about the people who generally tend toward either Type R or Type P. That may be obvious to Pascal, Gödel, and Heisenberg because they are accustomed to handling honest disagreements with other extremely bright and well-educated individuals, but it's a lesson that has eluded many zealots on both sides of the current environmental form of trench warfare.

Just as with Type I and Type II errors, neither of our new variations on the ancient theme of messing up is always better, or less bad, than the other. There are some circumstances under which Type P errors can be catastrophic, while other sets of conditions flip the situation entirely, making Type R mistakes the cause of disaster. Similarly, depending on the context, either breed of boo-boo can also be virtually harmless. Both Type R and Type P errors have the potential to range all the way from shoulder-shrugging to earth-shaking...but they take their most extreme form under very different scenarios. It is a matter of the utmost importance to have some way of distinguishing between these categories of outcome-shifting circumstances. Then we can focus directly and selectively on the more dangerous variety of error under those specific conditions.

The danger from making a Type R error is generally *acceptable* (and thus we should probably take action) when most or all of these conditions are *in effect*: (1) A correct, winning, active wager is very likely; (2) a correct, winning, active wager would bring us very large rewards; (3) an incorrect, losing, active wager is very unlikely; and/or (4) an incorrect, losing, active wager would not cost us an unacceptably large amount. That general category of situation tends to be a wake-up call, a bugle loudly blowing a clarion to-arms for the cavalry to charge into action. On the other hand, when most or all of these circumstances are *not* present, it is generally *unacceptable* to get actively involved, invest/gamble sizable amounts of resources, and take the risk of committing a Type R error. That breed of situation generally calls for us to calm down, enjoy some cheesecake, and hold our horses...or at least the remote control for our television.

Now look at the reverse situation. The risk of making Type P errors is generally *acceptable* (and thus it's okay *not* to intervene actively) when most or all of the following conditions are *present*: (1) A correct, winning wager if we actually take action and get involved is extremely unlikely; (2) such a correct, winning wager would bring us only small or moderate rewards; (3) an incorrect, losing wager if we take action is highly probable; and/or (4) such an incorrect, losing wager would cost us an unacceptably enormous amount. The size of the potential win or loss can be measured in terms of money alone, or (more realistically and accurately) money in conjunction with time, effort, inconvenience, quality of life impacts, positive and negative side effects, and other factors associated with the type of activity under consideration. Conversely, when most or all of these circumstances are *not* present, it is generally *unacceptable* to stay passively on the sidelines and take the risk of committing a Type P error. Under those circumstances, we are better off getting actively involved and launching some form of intervention.

The two sets of factors are mirror images of each other, That is, the combination of circumstances that tends to render it acceptable for us to risk committing a Type P error is the same collection that makes exposing ourselves to the risk of a Type R error unacceptable, and vice versa. This is not surprising, because a similar form of complementarity holds true for Type I and Type II errors; when we try to reduce Type I errors, we very often increase the incidence of Type II, and vice versa. Our gambling example is a bit different, though. It isn't only that we make Type R errors more probable by cracking down on Type P errors. It's that, but it's also more than that. Situations in which it's rational to accept the risk of Type R errors are the polar opposite of circumstances in which it's wise to assume the dangers inherent in Type P mistakes. The two types of gambling blunders are opposing sides of the same coin.

This makes sense, because when we are faced with an either/or decision (bet or don't bet, take action or stay out of it, buy insurance or not) any choice that embraces one option necessarily also rejects the other . At the poker table, we can either place a bet large enough to stay in the pot, or we can toss our cards into the muck. In situations of this type, there's no option to hedge our bets. An attempt to hedge, maybe by investing only in small pots or limiting ourselves to a moderate level of involvement, is really just a modified form of not betting at all, because a decision to act implies commitment to a level of action sufficient to fit the circumstances.

Any time we bet into a big—or potentially big—pot, we accept the risk of a Type R error while eliminating the chance of a Type P error at that same stage. That is, we take on the dangers associated with being overly aggressive (such as losing all we have in just that one hand) while we dodge any chance that we will be bitten by the reverse risk (such as missing out on playing the single hand that could have won an immense amount of chips for us). The flipside of the same coin is that we also embrace a risk—the risk that we are committing a Type P error—every single time we refrain from placing a bet large enough to keep us in a given pot…pre-flop, as well as after the flop, the turn, and the river. We implicitly choose to accept the chance that, by not betting aggressively, we might miss out on collecting a huge reward, in favor of *not* risking the immediate loss of all our chips. In taking our chances on a Type P error rather than Type R, we close the door to an instantaneous ruin, while also closing the door to an instantaneous windfall. Info-gap theory would identify this as abandonment of the opportuneness function. Either outcome can in fact change everything, but in fundamentally different ways.

I think this discussion of Type R and Type P errors helps to illustrate the oversimplification built into the motto "Better safe than sorry." Type R and Type P errors demonstrate that there is more than one approach to being "safe" and more than one way to end up being "sorry." I know. I've tried them both, many times. People who gravitate toward accepting Type P errors and avoiding Type R when confronting gradual, long-fused potential threats tend to believe in safety, in perhaps its most obvious sense—maximum reduction of the immediate and near-term probability of catastrophic loss from hasty, ill-conceived action. You can't win if you don't bet, but you also can't *immediately* lose all your chips either.

But this viewpoint, taken to an extreme, reflects a safety concept that is an illusion. Undue passivity is safe in the same deceptive and hollow sense as stuffing all of our cash into a basement mattress rather than investing it in some enterprise that might fail, or depositing it in a bank that might collapse. Inside our mattress, our money won't vanish in some corporate implosion or bank-busting Great Depression II, but it also will gradually lose its value to escalating inflation. Little by little with the remorseless passage of time, our nest egg will, inevitably, shrink and wither away to a pitiful little pile of impotent purchasing power rather than hatch and grow. No mattress ever invented pays interest that keeps pace with inflation. Plus, there's always some remote chance of a fire or burglary or flood, or some other form of unforeseen Black Swan peril wiping out our carefully-stuffed mattress. That's the hidden *absence* of safety and a subtle, usually slow-acting form of becoming sorry that comes from excessive reluctance to accept Type R error.

But activist, take-charge firebrands unavoidably also accept their own brand of being unsafe and ending up sorry. Can-do/will-do people often step in a lot of doo-doo. They prefer Type R errors over Type P, reflecting their affinity for active engagement in events they consider potentially important. This makes them vulnerable to the more obvious and immediate form of danger: aggressively investing large amounts of resources in a full-scale direct confrontation of a given risky, but slow-moving and long-term, situation. By pouring it on and pouring it in, they place vast amounts of their money, quality of life, time, work effort, and other valuable assets on the line, and accept the chance of losing it all. That is being unsafe in its most immediate, maxed-out, and glaringly blatant form. Clearly, when circumstances jerk the rug out from under such pull-out-all-the-stops commitment, it is indeed a sorry day. There's no way of not seeing it. The bad news is big, loud, and sudden. But unlike the unsafe and sorry consequences risked by those who prefer Type P error, Type R bad days can seem a little less bitter because, well, at least we struck out swinging instead of with the bat resting on our shoulder.

How, then, might Pascal use this new concept of Type R and Type P error to work through the climate decision matrix? With the help of Gödel and his familiarity with the Type I/Type II idea, Pascal could have added these ideas to his Wager and taken advantage of them to find an avenue to fresh views of our decision matrix. Let's try to imagine what shape this might take.

Pascal and Gödel would instantly grasp that the climate change decision is a case where it makes a big difference whether we are more likely to commit a Type R or a Type P error. Some situations are toss-ups as to which type of error might occur; but this isn't one of them. The reason is rooted in the magnitude and scope of the intervention necessary to even possibily reverse global climatic trends. We already recognized this when talking about the ways the climate change decision matrix differs from the hotspots matrix: The consequences of ineffective or unnecessary intervention are far worse in the climate change context. Because it is so much less certain that even an all-out war on climate change would solve any climatic problem, and because the wide spectrum of costs and losses we would incur would be so much more severe, it is far more important to be wary of Type R errors. The harm in being proactive but misguided, and betting big on needless and/or impossible climate intervention, is vastly worse than just a matter of unused insurance.

When it costs us a relatively modest, finite amount of money, inconvenience, life changes, and lost opportunities to take a precaution that has a good chance of protecting us from a possible huge downfall, it is often rational and prudent to take that precaution and accept the risk of a Type R error. Pascal would understand that, as an action-driving, cost/benefit calculation, this situation is closely akin to the original form of his Wager. But Pascal would also certainly appreciate the significant difference between that situation and the one we find in the climate change matrix. The possible results in the climate matrix itself reflect this. The shocking magnitude and gravity of a Type R error show up as an extraordinarily exorbitant form of insurance, so crushingly expensive it would typically be reserved for only the most desperate, last-ditch, dire-fire emergencies. Here's a major reason why Pascal's Wager in its original form pointed toward acceptance of Type R error and making the leap of faith: It's the comparatively slight, nearly negligible cost or downside potential inherent in choosing the positive action alternative. Not to overly trivialize the momentous, but one could look at the classic formulation of Pascal's Wager as an example of "Might as well, what have I got to lose?" decision-making.

To be sure, legitimate circumstances exist in which a dreadful long-term crisis warrants our intervention, even if our attempts at rescue may be dangerous, unlikely to succeed, and highly costly. Pascal would not deny this. But even those slow-motion cliff-hangers are not easy, shoulder-shrugging, off-hand cases where the action alternative is so obviously worth the risk that it seems foolish not to embrace it. If we ask ourselves, "What have I got to lose?" by launching an all-out climate intervention campaign, we might come up with a wide range of answers; but "Not much" wouldn't be among them. With probable aggregate costs in the trillions of dollars, plus profound life changes and economic upheaval, climate intervention is akin to an all-in bet in poker. It might be the best option under some circumstances, and if it pays off it could yield rich rewards, but it also places enormous value in jeopardy. A Type R error in this type of situation is a matter of immense consequences on both potential rewards and potential losses. Therefore our analysis of the decision matrix must take them both fully into account.

Looked at from another angle, there are formidable reasons to prefer the risk of Type P error in a situation like the climate change dilemma. This flows from the substantial rewards we can reap from *not* intervening, and remaining out of the game, when we know it will cost us a staggering price just to get into the action...and with no guarantee that we will actually win, even if we pay that price. As a generalization, Type P errors tend to be the preferable form of risk in situations very much like what we find in the climate change matrix—very high cost of betting/intervening, low or dubious chance of success from getting involved, and significant doubt as to the existence of a genuine and serious threat. Pascal, with Gödel at his side, would acknowledge it is prudent to require a significantly larger body of evidence—a higher burden of proof—before we decide to bet into the teeth of the risk of Type R error under these circumstances. We can reasonably demand much more conclusive evidence that our intervention is both truly necessary, and would in fact be effective, than we'd be willing to accept in the low-cost/high-success-rate "might as well" variety of situations.

We could think of this as a principle related to the nature of Type P and Type R errors. I've explained that there tends to be some combinations of factors that favor inaction, and thus acceptance of Type P error risk, while other clusters of conditions push us toward action, with the associated assumption of Type R risk. These are generalizations. They are tendencies. They are not inflexible, stone-carved, capital-C Commandments. *But* before we make a decision that swims against the tide, the decision matrix cautions us not to do so without an extra measure of evidence and assurance. In situations that usually favor acceptance of Type P errors more than Type R, we would be wise to insist on a higher burden of proof—a greater quantum of credible evidence—before we decide to make the contrarian decision and embrace the risk of Type R error. A more restrictive threshold is therefore usually appropriate where the context trends in the direction of accepting Type P risk. That would guard against overly aggressive intervention where circumstances don't warrant it.

And, where the context mostly points toward preferring Type R risk, a more sensitive threshold is the norm, to minimize the risk of failing to intervene when intervention is justified. We would be wise to demand a lot more evidence than usual that the status quo is the way to go when our warning lights are all flashing and the alarm bells are clanging out an alert. That is, when most indications point toward a true, severe emergency that we could likely avert at low (or at least acceptable) cost and risk, we'd be smart to take action unless a large amount of powerful evidence cautioned us not to swim against that current and stay where we are.

Along these lines, it is logically defensible, Pascal might argue, to require more evidence, and more objectively reliable evidence, that our decision is appropriate in situations where we are considering making a choice that runs *counter to* the natural arrangement of

Type R and Type P errors. If the normal preference under a given set of conditions would be in favor of running the risk of a Type R error, we should demand a stricter burden of proof from anyone who urges us to stay out of the action. Conversely, under conditions where it is usually deemed reasonable to accept the risk of committing a Type P error, we are justified in insisting on a stronger evidentiary case from anyone who would persuade us to get involved notwithstanding the norms.

People who wouldn't know Gödel from Google or Heisenberg from Hannah Montana can appreciate the reasoned approach to uncertainty embodied in the World Wager and its more tailored variations (such as the climate wager and the hotspots wager). Insurance is part of the common experience of many adults all over the world, and it's a familiar setting in which we confront the uncertainties and probabilities of life, taking into account the potential risks and rewards under the various possibilities. We don't wait to buy life insurance until we know we are seriously ill; in fact, if we delay our insurance-buying decision that long, we may find it is no longer an option at any price. Instead, we have to plan ahead, and figure out a reasonable path through a thicket of unknowns, much like those that haunt the hotspots wager. How likely are we to suffer the misfortune in question (whether a house-destroying fire, or our own death, or some disabling physical problem) within the next year? What would it cost us now to purchase insurance sufficient to ameliorate to a tolerable level the damage from such a misfortune? What would happen to us and/or our loved ones if we opt not to buy insurance now, but then disaster strikes? What else could we do with the money we'd save on insurance premiums if we decide not to obtain insurance? If we postpone our insurance purchase until we see clear symptoms, what is the probability that we would be able to acquire adequate insurance later, and at an affordable price, if we then decide we really do need it?

These issues, laden with unknowns, are dealt with at some level by everyone who makes a decision on paying for insurance. They have obvious parallels to the factors underlying the World Wager. And, significantly, millions of people decide every year that it is a prudent choice to buy insurance on their life or home. They make this decision despite the fact that, by any objective measure, there is a very low probability that they will actually need the insurance in any particular here-and-now year. They grasp the hard truth that, somewhere down the line, their luck may run out, and they need to plan ahead for that sad day if they want to be responsible stewards of their own lives and those of their loved ones. They decide it's not a sound gamble to bet their life and their family on their luck staying good indefinitely.

This is the Precautionary Principle and the World Wager at work in vast numbers of normal lives, in practical and important personal decisions, replicated annually on a worldwide scale. Millions of times each year, people who have never heard of the Precautionary Principle nonetheless understand its "better safe than sorry" "might as well" essence. They know that it is sometimes a reasonable choice to spend a sizable but limited amount of their own money every year to be prepared for a potentially devastating tragedy, even though the probability of such a tragedy actually befalling them that year is far less than 50 percent. Indeed, the probability of a fire actually destroying our house in the next 12 months must be vanishingly small, yet hosts of us still consider our fire insurance premiums a worthwhile precaution, year after fire-free year. We make that choice because somehow we intuitively appreciate the underlying point of Pascal's Wager and my World Wager variant. We understand that when we face an important issue with incomplete information, we must factor in the possible best-case and worst-case scenarios under each set of available decisions, along with the odds of each one. We know that the amount it takes to purchase replacement-value fire insurance on our home is overwhelmingly less than the money we would lose if our house burned down and we had no insurance. We judge that it isn't worth the gamble to risk the huge cost of replacing our entire home merely to save a comparative pittance in insurance premiums, irrespective of the slight probability of disaster. We may even conclude that to do otherwise would be indicative of a gambling problem.

In our ordinary experience, what do we mean when we say someone has a gambling problem? Usually, we reserve this harsh judgment for cases of personal irresponsibility and compulsion beyond any normal flirtation with games of chance. If we see substantial indications that a person has most or all of these symptoms, we see the need for expert intervention:

- Gambling with resources he or she can't afford to lose.
- Risking resources that are needed for his or her family, especially children.
- Immersion in a game he or she doesn't understand, in terms of the probabilities, rules, limitations on our ability to control outcomes, and risk/reward ratios.
- Obsession with a game at which he or she lacks the skill to win regularly.
- Fixation on a game where luck or other factors beyond his or her control are the predominant or sole influence on the outcome.
- Failure to limit gambling to a level that does not seriously impair important responsibilities and life functions.
- Inability or unwillingness to stop risking more resources when the results are bad.
- Betting based on emotion (e.g., fear, anger, impatience, frustration, anxiety) rather than calm, rational, and logical assessment of the odds, risks, and rewards.
- Elevation of the game above the most important basic life activities (work, relationships, fun).

Now put this book down for a few minutes and just let yourself think. Think about the way we are handling mass extinction, climate change, and any of the other momentous environmental challenges in the world today. Gather all the reliable facts you can from all reasonable sources. Identify the key unknowns, gaps in information, and variables. Think about how the World Wager and decision matrix could be used to weigh all the options, all the opportunities and perils, in light of the probabilities. Then ask yourself whether we are handling the situation in an intelligent, prudent, responsible manner. Or do we have a gambling problem here? What is your answer to that question, based on all you know?

Let's return briefly to one of the crowning insights from Gödel for some clues as to the answer. His incompleteness theorems reveal that, within any system of study, more things are true than can be proved. Because truth is a more powerful principle than provability, uncertainty is inevitable as we try to play by the internal rules of any field. We might *feel* that some ideas are true, and our intuition may have a very strong hold on our beliefs, but still we will be unable to prove, definitively and beyond dispute, every single notion. To prove some points, we would necessarily need to rely on some other source of information, beyond the matrix within which we are working. In this sense, the systems with which we deal are incomplete. They are also indefinite. More aspects of reality within every system are true than we can ever prove to be true without departing from the system itself.

We are left with imperfect but useful tools such as the Precautionary Principle and the World Wager/decision matrix. When we rely on these devices, we don't surrender our rigor or throw our reason onto the altar of guesswork. Quite the contrary, we accept the inherent limitations that attach to our ability to understand or measure reality. If this is in some sense a leap of faith, it is more akin to a small, deliberate, well-reasoned step than a blind, headlong jump into a dark chasm. Faith of this nature is not illogical, but rather the most logical way of handling the uncertainty and incompleteness built into the universe. It allows us to make prudent decisions that take into account rather than ignore the probabilities associated with all significant holes in our information foundation. If our system is incomplete, then we are better off being honest about it and doing our best to discern the likelihood of the various alternatives than to pretend that we know everything worth knowing and that all is rock-certain.

This is a key point. Gödel might point out that many people make their most vital life-changing decisions while under the burden of a self-imposed evidentiary equivalent of a starvation diet. We behave as if we were two-dimensional characters from *Flatland*. Like Flatlanders, we don't see the truth of some facts or principles that a three-dimensional person would recognize as being very nearby, but on the other side of a thin, flimsy line. A Flatlander wouldn't be able to see anything on the other side of that line, because in a planar world, any line is as insurmountable a barrier as the highest mountain range is to us. But someone from Spaceland would have a more complete perspective and would easily find the "hidden" truth. Looking down from above, with the immeasurable benefit of a third dimension, Spacelanders can see the truths of things beyond the limits of a Flatlander's paper-thin methods of proof.

Gödel's first incompleteness theorem is analogous to the limitations legitimately inherent in a Flatlander's two-dimensional perspective. When your perspective is limited, some things can't be proved even though they're true. There are some limits on our ability to prove, built into our system. But Gödel would laugh his high-pitched laugh at some of us, because we voluntarily deprive ourselves of information and ideas that could enable us to prove the truth of more (but not all) true principles in our world. We forsake the advantages of three dimensions and limit ourselves to only two, as if we were Flatlanders. We constrict our perspective artificially and needlessly. In doing so, we throw away our Spacelander advantages and join the

Flat Earth Society. We deliberately surrender some of our ability to test, to prove, and to analyze. We take the limitations on provability that come with any incomplete system and unilaterally, foolishly multiply them. We herd our horde of possible Type R and Type P errors together, crowd all of them into a nice, cozy, dimly-lit romantic setting, and allow them to have lots and lots of fuzzy children.

It's one thing to be born a two-dimensional denizen of Flatland. There's nothing we can do to eliminate the innate limitations hardwired into any system, as Gödel would remind us. But it's another, and far worse, thing entirely to tighten a tourniquet ourselves around our own informational pipeline. When we refuse to consider, objectively and rationally, the merits and weaknesses of all pertinent bits of evidence, we expand the impoverishing effects of Gödel's theorems on our decision-making capabilities. By preemptively ruling out fair and full consideration of much of the evidence, we take a metaphorical truckload of facts from the true-and-provable pile and dump them instead into the leap-of-faith heap. As I argued in Chapter 5, this isn't a necessary limitation inexorably restricting our ability to discern truth from error—this has nothing to do with Gödel's incompleteness theorem. Instead, it is the incompleteness of our own honesty, intellect, open-mindedness, preparation, education, and integrity that compels us to transfer provable truths from the realm of scientific rigor to the clouds of quasi-religious dogma where everything must be taken on faith.

Heisenberg might look at what we've done, and what we've failed to do, and shake his head in wonder. Why, he might ask, do intelligent people relinquish tools and materials properly available to us and needlessly deprive ourselves of essential ingredients in our experimentation? Gödel would add that the rational approach is to maximize our access to pertinent facts and principles, not to discard some of them based on ignorance or prejudice. Pascal, too, would marvel that modern people, who make such a show of deprecating religion and elevating science, would actually anathematize their opponents and rule their evidence out of bounds, based not on sound and objective scientific principles but on something akin to a matter of dogmatic faith. He'd be appalled that there would be so many people around today who seem so eager to emigrate to Flatland.

It is in the evidence-screening phase of the decision-making process where we see the worst cases of Voluntary Flatland Syndrome (VFS). Oh, you've never heard of VFS? Well, that's because I just made it up. VFS is the situation where we—either deliberately or carelessly—exclude from our choice-making process some important, relevant, credible evidence. VFS amounts to a very serious self-inflicted wound, one which may well be fatal in some cases. In a complex situation where it is highly challenging to determine what studies, theories, and experimental results are within the ballpark of legitimacy, we give ourselves a bad and needless case of shooting ourselves in the foot (or the head) when we multiply those problems. But that's what we do when we unjustifiably rule out consideration of ideas and data from the outset. We take a bad hand and misplay it until it's downright disastrous. We take the inherent limitations on precision and provability that are unavoidably standard equipment in our system and we knowingly expand and reinforce those limitations. It's like deciding to do all our driving, all the time, in reverse while looking only at our rear-view mirror. To a considerable degree we are inextricably imbedded in a real version of Flatland—hemmed in by built-in barriers to our understanding and perception—with all the artificial ceilings and walls that come along with those systemic constraints. Rather than add to them, it would seem preferable to do all we can to minimize the restrictions on our ability to learn, to perceive, and to comprehend reality...from the vantage point of as many dimensions and angles as we can incorporate.

Recall my examples of a possible inoculation-autism link (from Chapter 8) and the question of an element of skill in playing the lottery (from Chapter 9). I made the judgment call to deem at least *somewhat* credible the marginally supported evidence of a connection between inoculation and autism, so that it could be included in the mix of all relevant and minimally competent factors to be weighed in a decision on inoculation. I made the opposite call on the existence of a skill component in lottery gambling, finding insufficient indicia of reliability, theoretical validity, rigorous methodology, and observational objectivity underlying the anecdotal reports of skill as a determinant of success among lottery players. This threshold ruling relegated playing the lottery to the luck-locked status of a roll of the dice rather than the skill-driven play of a poker hand. By preemptively ruling out evidence of a skill factor in lotteries, I effectively rendered it impossible for lotteries to be viewed as more similar to poker than to craps or bingo...and that, in turn, shifted the entire question into the category of purely random coin-flip gambling rather than rational, capabilities-based investment in our own competency. *This peremptory exclusion of evidence is not something we should do lightly.* As with any judgment calls, both of these choices were to some extent matters of subjective opinion, with ample room for my own arbitrariness, personal predisposition, educational background, unsupported assumption, frame-of-reference bias, and experience-based capriciousness to influence the outcome.

Obviously, I am a big fan of my World Wager and decision matrix. Why not? I created them, stealing shamelessly from the best minds of all time to craft them into the most accurate and powerful tools available for making Earth-shaking decisions. But they are helplessly and totally dependent on the information we feed them. If we thoughtlessly starve them or offer them only an unbalanced diet, they will vomit out a raging torrent of Type R and Type P errors like a veritable perpetual-emotion machine. As with the most powerful computer ever designed, the unavoidable and absolute limiting factor on their effectiveness is the quantity and quality of the input they receive from us. They can't give us decisions any better than what can be derived from the facts and principles we provide for them to work with. If we put garbage in, we shouldn't be too shocked when we receive results that share a number of features in common with garbage as well. Even the best decisional devices can't perform alchemy. Pity.

We seem to be relying on alchemy, or some form of Midas touch, to rescue us from ourselves as we make poorly informed decisions that can determine the fate of the Earth. On climate change, biodiversity loss, and other matters of acute significance, many of us are falling victim to VFS. We are preemptively excluding some evidence, and reflexively overvaluing other evidence, without full and fair evaluation. In doing this, we are failing to fuel our decisional machinery, starving it of crucial foundational components. Before we have an opportunity to see where the World Wager and decision matrix would take us if we gave them a full tank to go on, we're letting them run out of gas. You can't save the world when your engine has nothing but vapor to work with.

It's difficult enough under the best of circumstances to decide what evidence is credible, corroborated, reliable, theoretically sound, and logically consistent enough that we should include it in our mix of relevant factors. In fields where even subject-matter experts disagree vehemently, and where the underlying principles are complex and hard to grasp without years of advanced education, all of us face a daunting task on these threshold options. We are, to a great degree, in the position of Flatlanders. We can't see things someone with a higher perspective might find ridiculously obvious. We misperceive other things because of our circumstances, the obstacles in our path, our cultural context, our social upbringing, the experiences we've had, and our schooling...regardless of how hard we try to see the truth or how sincerely we desire to do what's right. Like a Flatlander trapped inside a hollow square, our situation renders it impossible to break free, unless somehow we can gain the added dimension and perspective to let us see out of the box. We are trained and conditioned during our whole lives to think and reason and perceive certain things in a certain way, and it requires a great deal of effort and determination to move beyond that default option when the chips are down. But if we don't, our choices and our decisions are made for us in advance, and all of our chips may soon vanish, before we even begin.

Many of us fail before we start. We take our culture and turn it into a cult. Rather than doing all we can to loosen the bonds of our background to see farther and with greater clarity, we surrender to our surroundings. We defer to one or more handy and comfortable "experts" to whom we abdicate our ability and our right to think independently. And we follow our own familiar herd, going along with whatever "everybody" thinks about the issue within our tight circle of friends, colleagues, and family. As long as we stay within our particular comfort zone and home turf, this is a very easy and undemanding course of inaction. We let others do the paddling and the steering for us, and we effortlessly go with the flow, laughing at anyone stupid enough to swim against the strong current of our preferred brand of truth. But this is what VFS does to us, if we allow it. We voluntarily give up our access to additional dimensions and differing perspectives. We flat-line our brain waves. With a self-induced, utterly straight and uncomplicated horizontal EEG, we slide sideways into our nearby two-dimensional cell, drag the door firmly closed behind us, and remain for a life-sentence, just as boxed in as if we were born into Flatland and were forced to be there.

It is a matter of the highest importance that we elevate the level of discussion concerning these environmental threats. I believe many people in all walks of life—and maybe even some in government—in all nations will respond positively if the issue is properly framed. It is counterproductive to make unsupportable claims of scientific certitude. I have tried to show how the inevitable uncertainties can be acknowledged and dealt with prudently and realistically, much like the doubts and unknowns with which we all must deal in our everyday lives. You need not be a genius on the level of a Pascal, Gödel, or Heisenberg to grasp the relevance of their breathtaking insights to today's most monumental issues. You just have to be open to a frank conversation about the role of uncertainty in all of life's largest decisions...whether on a global scale or one person at a time. Only in the Disney version of fairytales are you absolutely 100 percent guaranteed to live happily ever after. Of that, at least, there is no doubt.

When you're gambling with what you absolutely can't afford to lose, you can't afford to push all-in blind, in the dark. If there ever were a betting decision where you want every light on, every distraction shut out, and expert advice from the best possible minds, this is it. We've been compelled, against our will, to bet the Earth in the highest of all high-stakes poker games. What's the best we can do with the cards we've been dealt? Is this the time for us to play a hunch, rely on a lucky charm, flip a coin, or consult a fortune cookie? Is this a good opportunity for us to close our eyes to credible evidence and reasonable people with messages we just don't want to consider? Or is the biggest bet of all time worth the effort to be as informed, prepared, logical, rational, and accurate as we can possibly be?

Gambling isn't an absolute wrong or right. Under many circumstances in our daily lives, it's unavoidable. We're constantly forced to take chances, weigh the odds, and pick among options in the dim light. We might call this a necessary evil. But evil or not, it's necessary in the sense that there's no way to stand aside and remain a spectator in all these everyday games of chance. Our chips, our future, and our lives are at risk, and we are in the game whether we like it or not.

But what about compulsive gambling, irresponsible gambling, gambling with resources we can't afford to lose? That's the type of risk-taking that ruins lives. When we bet with our family's rent money or our kids' college fund, we're doing something deep-down different, and worse, than a little casual low-stakes play. We're playing for keeps with something we can't afford to lose, and isn't really even entirely ours. Betting of that kind isn't necessary, but it is evil. It's playing with fire in the middle of a dynamite factory. If you risk all those chips with your loved ones' lives on the line, and the lives of all their children and grandchildren that might follow, you'd better be right. Reckless wagers now can wreck the future for generation upon generation to come, in a devastating domino collapse of dreams for even unborn descendants whose lives will be fundamentally worse because of what we do and fail to do today. If you're not going to play it safe in that situation, sorry isn't even close to good enough.

What are we doing, then, when we gamble with the Earth? We're betting with resources we can never win back once they're lost. We're taking chances with chips that, in part, belong to each of us, but in whole to none of us. We're gambling with the rent money for the next month, the next year, and the next few thousand years. And we can't throw our hands toward the sky and beg off to watch from the rail. We have to play this game. It comes with the package deal for everyone alive. The exit doors don't open, and our chips *are* at risk. So what's our best bet?

I wrote this book to offer a way of determining our *best* bet. There's no such thing as a safe bet here. There are no sure things in this casino, or any other. In fact, there are distractions, temptations, hidden pitfalls, and deceptively pleasant illusions on all sides. The lights are low, the noise is loud, the intoxicants flow freely, the neon flashes, and the smoke is thick. It's terrible to be trapped in a high-stakes game of chance under these conditions. Complain to the management if you'd like, but it won't do any good.

The best we can do—the best any of us can do—is to approach this monumental wager with our eyes, and our minds, wide open. But this is where we've failed so miserably for so long. The most basic, fundamental question any gambler must ask is this: *What game am I playing?*

In many ways we don't even know, because we've never really checked. We don't know the rules, or the practical limitations of what's possible. For the moves we might possibly make, we haven't a clue as to their probability of bringing us win or ruin. We haven't done our homework on how much we stand to gain or lose under various circumstances. We aren't even truly sure whether this game we're in has a component for our skill to make a difference, or whether we're completely at the mercy of Lady Luck disguised as Mother Nature. We don't know whether this high-roller game holding us hostage is closer to poker or bingo. And that's something you'd really prefer to know when your planet is riding on the result.

If this book is about anything, it's about learning and thinking. We must begin by learning what type of game we're playing with our planet. That means doing all we can to ask, and arrive at the best possible answer for, the Earth's most bedrock foundational questions. It means learning and thinking about what's truly at stake, what it takes to win, what the possibilities are, where the probabilities trend, and which way the horns of every dilemma point when we wade through all the smoke and fog. If we can identify the type of game we're in, think all our options through, gather all the relevant facts, and then honestly, logically, and rationally consider all the possible risks and rewards, odds and oddities, at least we won't be helpless fish in a sucker's game. We may not win the gamble, but at least we won't have thrown it all away in a rigged game like a drunken fool. Unavoidably, irreversibly, all of us are betting the Earth, right now. We owe it to the future to place those planet-sized bets wisely.

Let's consider one last cautionary story that vividly symbolizes the plight we face. Prometheus is one of the most heroic and tragic figures from ancient Greek mythology. To help humans, Prometheus stole fire from the King of the Gods, Zeus, and gave this life-enhancing power to the mortals. What ensued was an early illustration of the wry cliché, "No good deed goes unpunished." Zeus punished

Prometheus for his attempt to contribute to global warming by having him bound to a rock for all eternity. But that wasn't all. Zeus also arranged for a giant eagle to rip out and devour Prometheus's liver every day, which would then grow back so the eagle could repeat the gory torture the next day. Forever. According to some tellings of the myth, Zeus went on to send Pandora to the mortals, carrying a tempting enclosure, as further retribution for their bold arrogation of power. Prometheus' warnings went unheeded, and the unthinking mortals accepted Pandora's malevolent yet tantalizing gift. This ushered into the world an onslaught of evils, pains, and deadly diseases. What still remained inside Pandora's fateful container after the lid was again closed? Only hope.

We are like the tragic figures of ancient myth. We are compelled to wager our lives, all we love, and our very world in a cruel game of chance. Did we somehow, unknowingly, bring this torture upon ourselves, like a Prometheus for the end of time? In seizing the secret powers of energy and industry for us mortal humans, have we provoked the wrath of unseen and unlimited forces? The furies we have unleashed in the act of harnessing Nature's might for our service have now trapped us in an epic dilemma. We are left with no choice but to gamble, with all of our tomorrows at stake. And Nature's vengeance compels us to place our monumental wagers within a shroud of fog, with the choices and the chances only dimly and partially lit by a flickering lamp. How can we survive this merciless gamble? How can we solve the puzzle of this game and choose the bets that bring life? Our only hope, and the only hope of our world, is to tame the same darkness and mist that conceal the truth. It is only by piercing straight through the heart's core of that uncertainty that we might yet find our way to the hopeful light beyond this most deadly game of chance.

Footnotes

CHAPTER 1

1. Endangered Species Act, 16 U.S.C. §§1531-1544.
2. *See generally* Lear, Linda. RACHEL CARSON: WITNESS FOR NATURE, New York: Henry Holt (1997); Lytle, Mark Hamilton. THE GENTLE SUBVERSIVE: RACHEL CARSON, SILENT SPRING, AND THE RISE OF THE ENVIRONMENTAL MOVEMENT, New York: Oxford University Press (2007).
3. SILENT SPRING, Houghton Mifflin, 1962, Mariner Books, 2002. *Silent Spring* initially appeared serialized in three parts in the June 16, 23, and 30, 1962, issues of *The New Yorker* magazine.
4. *See generally* Craig E. Colten and Peter N. Skinner, THE ROAD TO LOVE CANAL: MANAGING INDUSTRIAL WASTE BEFORE EPA, University of Texas Press (1996).
5. Comprehensive Environmental Response, Compensation, and Liability Act (CERCLA), 42 U.S.C. §§9601-75, enacted on Dec. 11, 1980.
6. Toxic Substances Control Act (TSCA), 15 U.S.C. §§2601-92.
7. *See generally* Rachel Maines, ASBESTOS AND FIRE: TECHNOLOGICAL TRADE-OFFS AND THE BODY AT RISK, Rutgers University Press (2005).
8. *See generally* Larry G. Hansen and Larry W. Robertson (eds.), PCBs: HUMAN AND ENVIRONMENTAL DISPOSITION AND TOXICOLOGY, University of Illinois Press (2008).
9. *See generally* Edward A. Parson, PROTECTING THE OZONE LAYER: SCIENCE AND STRATEGY, Oxford University Press (2003).
10. Edward C. Krug and Charles R. Frink, Acid Rain on Acid Soil: A New Perspective. 221 Science 520-525 (1983); G.E. Likens and F. H. Bormann, Acid Rain: A Serious Regional Environmental Problem, 184 Science 1176-1179 (1978).
11. Howard Kurtz, Is Acid Rain a Tempest in News Media Teapot? Washington Post, January 14, 1991 at A3; Anne LaBastille, How Menacing is Acid Rain?, National Geographic 160(5): 652-681(1981).
12. See, e.g., Dallas Burtraw and Byron Swift, A New Standard of Performance: An Analysis of the Clean Air Act's Acid Rain Program, 26 Envtl. L. Rep. (Envtl. L. Inst.) 10411 (1996); Byron Swift, The Acid Rain Test, Envtl. F., May-June 1997, at 17.

13 *See generally* Murphy, Priscilla Coit. WHAT A BOOK CAN DO: THE PUBLICATION AND RECEPTION OF SILENT SPRING, Amherst: University of Massachusetts Press (2005).
14 Jeff Zeleny and Jackie Calmes, Obama, Assembling Team, Turns to the Economy, New York Times, Nov. 6, 2008.
15 Pete Harrison, Never waste a good crisis, Clinton says on climate, Reuters India, Mar 7, 2009 < http://www.reuters.com/article/idUSL6592864>.

CHAPTER 2

1 Michael Mann, Raymond Bradley, and Malcolm Hughes, Global-scale temperature patterns and climate forcing over the past six centuries, 392 NATURE 779 (1998), <http://www.caenvirothon.com/Resources/Mann,%20et%20al.%20Global%20scale%20temp%20patterns.pdf>; Mann et al., Testing the Fidelity of Methods Used in Proxy-Based Reconstructions of Past Climate, 18 JOURNAL OF CLIMATE 4097 (2005), <http://www.meteo.psu.edu/~mann/shared/articles/MRWA-JClimate05.pdf>; Mann et al, Proxy-based reconstructions of hemispheric and global surface temperature variations over the past two millennia, 105 PNAS 13252 (2008), <http://www.pnas.org/content/105/36/13252>
2 IPCC Third Assessment Report (2001) <http://www.grida.no/publications/other/ipcc_tar/>.
3 IPCC Fourth Assessment Report (2007) <http://www.ipcc.ch/ipccreports/ar4-wg1.htm>.
4 Stephen McIntyre and Ross McKitrick, Corrections to the Mann et. Al. (1998) Proxy Data Base and Northern Hemispheric Average Temperature Series, 14 ENERGY & ENVIRONMENT (2003), <http://www.uoguelph.ca/~rmckitri/research/MM03.pdf>
5 *See, e.g.,* Geoff Brumfiel, Academy affirms hockey-stick graph, 441 NATURE 1032 (29 June 2006); Board on Atmospheric Sciences and Climate, Surface Temperature Reconstructions for the Last 2,000 Years, National Academies Press (2006), <http://books.nap.edu/openbook.php?record_id=11676&page=R1>; Wegman Congressional Committee Report <http://republicans.energycommerce.house.gov/108/home/07142006_Wegman_Report.pdf>
6 Gerald Traufetter, Climatologists Baffled by Global Warming Time-Out, Der Spiegel (Nov. 30, 2009), < http://www.spiegel.de/international/world/0,1518,662092,00.html>.
7 NASA's Goddard Institute for Space Studies. < http://data.giss.nasa.gov/gistemp/graphs/Fig.A2.pdf>.
8 *Id.*
9 Paul Pearson and Martin Palmer, Atmospheric carbon dioxide concentrations over the past 60 million years, 406 NATURE 695-99 (17 Aug. 2000).

10   J.R. Petit et al, Climate and atmospheric history of the past 420,000 years from the Vostok ice core, Antarctica, 399 NATURE 429 (3 June 1999), <http://www.daycreek.com/dc/images/1999.pdf>.
11   EPA, Recent Climate Change: Atmospheric Changes, Climate Change Science Program <http://www.epa.gov/climatechange/science/recentac.html>.
12   Jones, P.D. and Moberg, A, Hemispheric and large-scale surface air temperature variations: An extensive revision and an update to 2001, 16 JOURNAL OF CLIMATE 206 (2003); Christy et al, Error estimates of version 5.0 of MSU/AMSU bulk atmospheric temperatures, 20 J. ATMOS. OCEANIC TECHNOL.613 (2003); Matthias Schabel, et al, Stable Long-Term Retrieval of Tropospheric Temperature Time Series from the Microwave Sounding Unit, PROCEEDINGS OF THE INTERNATIONAL GEOPHYSICS AND REMOTE SENSING SYMPOSIUM, 1845 (2002).
13   *See* Paul Pearson et al., Global Temperature Change, 103 PNAS (39) 14288-14293 (Sep. 26, 2006).
14   *See generally* Lisa Heinzerling, Climate Change in the Supreme Court, 38 ENVTL L. 1 (2008).
15   *See generally* Francesca Fonte, The Threat of Global Warming and the Roles of the EPA: EPA as an Independent Agency and as Advocate for the Environment and the Public?, 11 ALB. L. ENVTL. OUTLOOK 365 (2007).
16   Massachusetts v. EPA, 549 U.S. 497 (2007).
17   42 U.S.C. 7401-7671q.
18   *See generally* <http://www.epa.gov/climatechange/endangerment.html>
19   *Id.*
20   Naomi Oreskes, The Scientific Consensus on Climate Change, 306 SCI. 1686 (3 Dec. 2004).
21   Gerald Traufetter, Climatologists Baffled by Global Warming Time-Out, Der Spiegel (Nov. 30, 2009), < http://www.spiegel.de/international/world/0,1518,662092,00.html>.
22   *See generally* Marcel Leroux, GLOBAL WARMING- MYTH OR REALITY?: THE ERRING WAYS OF CLIMATOLOGY (Springer, 2005): Christopher Horner, THE POLITICALLY INCORRECT GUIDE TO GLOBAL WARMING (AND ENVIRONMENTALISM), (Regnery, 2007) (critiquing several causes to which environmentalists have tried to rouse public interest).
23   *Climate Change: Basic Information*, Environmental Protection Agency Online, <http://www.epa.gov/climatechange/basicinfo.html, (last accessed Sept. 1, 2009)>.
24   Nordhaus, William, A QUESTION OF BALANCE, 2, (Yale University Press 2008).
25   Viscount Monckton of Brenchley, *An Open Letter from the Viscount Monkton of Brenchley to Sen. John McCain about Climate Science and Policy*, American Thinker, <http://www.americanthinker.com/2008/10/an_open_letter_from_the_viscou_1.html> (Oct. 18, 2008).
26   *Id.*

27 Carlin, Dr. Alan, *Proposed NCEE Comments on Draft Technical Support Document for Endangerment Analysis for Greenhouse Gas Emissions under the Clean Air Act*, National Center for Environmental Economics, 70 (Mar. 2009).
28 *Id.* at <http://www.americanthinker.com/2008/10/an_open_letter_from_the_viscou_1.html>.
29 *Id.* at Carlin, National Center for Environmental Economics, 67.
30 *Id.* at 69.
31 *Id.*
32 *Id.* at 11.
33 *Id.* at <http://www.telegraph.co.uk/comment/columnists/christopherbooker/3982101/2008-was-the-year-man-made-global-warming-was-disproved.html>.
34 Murdock, Deroy, *Whatever Happened to Global Warming?*, National Review Online, <http://article.nationalreview.com/?q=NGYzMzA2MzEzZTI4YjAxOTZhMGY4N2YwOTVmZWIzOTg>, (Dec. 23, 2008).
35 Coulter, Ann, *Our Mistake—Keep Polluting*, Jewish World Review, <http://www.jewishworldreview.com /cols/coulter090800.asp>, (Sept. 8, 2000).
36 *Id.* at <http://www.vancouversun.com/news/Global+warming+religion+First+World+urban+elites/1835847/story.html>.
37 Coulter, Ann, *Gore's Global Warming Religion*, Human Events, <http://www.humanevents.com /article.php?id=19927>, (Mar. 21, 2007).
38 *Id.* at <http://www.vancouversun.com/news/Global+warming+religion+First+World+urban+elites/1835847/story.html>.
39 Gonzalez, Mike, *An Inconvenient Voice: Dr. Alan Carlin*, The Heritage Foundation Online, <http://blog.heritage.org/2009/06/29/an-inconvenient-voice-dr-alan-carlin/>, (Jun. 29, 2009).
40 *Id.*
41 *Id.*
42 Heartland Institute Staff, *New York Global Warming Conference Considers 'Manhattan Declaration,'* The Heartland Institute, <http://www.heartland.org/policybot/results/22866/New_York_Global_Warming_Conference_Considers_Manhattan_Declaration.html>, (Mar. 4, 2008).
43 *Id.* at Carlin, National Center for Environmental Economics, 3.
44 *Id.* at 5.
45 *Id.*
46 *See* <http://data.giss.nasa.gov/gistemp/graphs>; <http://www.ncdc.noaa.gov/oa/climate/globtemp.html>; <http://www.epa.gov/climatechange/science/recenttc.html>.
47 *Id.* at Carlin, National Center for Environmental Economics, 4.
48 *Id.*
49 Telephone Interview with Lord Monckton of Brenchley, (Feb. 12, 2009).

50 Hall, Ed, U.S. National Debt Clock, <http://www.brillig.com/debt_clock/> (last updated Sept. 8, 2009).

51 Nordhaus at 77.

52 The Editors of the National Review Online, *Getting Warmer*, The National Review Online, <http://article.nationalreview.com/?q=NzNjY2VmMzAyN2ZkNmNhZG RkYmIxOWI5ZTZmZDg2Y2E>, (Dec. 12, 2008).

53 Telephone Interview with Lord Monckton of Brenchley, (Feb. 12, 2009).

54 *See generally* National Academy of Sciences Committee on the Science of Climate Change, CLIMATE CHANGE AND SCIENCE: AN ANALYSIS OF SOME KEY QUESTIONS (National Academy Press, 2001); American Meteorological Society, 84 BULL. AM. METOROL. SOC. 508 (2003); American Geophysical Union, 84 EOS 574 (2003).

55 *See, e.g.,* Joint science academies' statement: Global response to climate change (June, 2005), available at < http://royalsociety.org/displaypagedoc.asp?id=20742>.

56 The Convention was adopted on May 9, 1992, and entered into force on March 21, 1994. The Kyoto Protocol builds on the foundation established by this Convention.

57 *See generally* Lawrence Solomon, THE DENIERS: THE WORLD RENOWNED SCIENTISTS WHO STOOD UP AGAINST GLOBAL WARMING HYSTERIA, POLITICAL PERSECUTION, AND FRAUD, AND THOSE WHO ARE TOO FEARFUL TO DO SO, Richard Vigilante Books (2008); Chris Mooney, STORM WORLD: HURRICANES, POLITICS, AND THE BATTLE OVER GLOBAL WARMING (Harcourt, 2007); S. van den Hove, M. Le Menestral, H.C. de Bettignies, 2 CLIMATE POLICY (1) 3 (2003).

58 *See* Robert Mendick, "Climategate: University of East Anglia U-turn in climate change row," Telegraph, Nov. 28, 2009. <http://www.telegraph.co.uk/earth/copenhagen-climate-change-confe/6678469/Climategate-University-of-East-Anglia-U-turn-in-climate-change-row.html>

59 *See* Christopher Booker, "Climate Change: This is the worst scientific scandal of our generation," Telegraph, Nov. 28, 2009. <http://www.telegraph.co.uk/comment/columnists/christopherbooker/6679082/Climate-change-this-is-the-worst-scientific-scandal-of-our-generation.html>

60 *See generally* Steven Mosher and Thomas Fuller, CLIMATEGATE: THE CRUTAPE LETTERS (Volume 1), CreateSpace (2010); A.W. Montford, THE HOCKEY STICK ILLUSION: CLIMATEGATE AND THE CORRUPTION OF SCIENCE, Independent Minds (2010).

61 *See* Keith Johnson, "Hacked: Sensitive Documents Lifted from Hadley Climate Center," Wall Street Journal, Nov. 20, 2009.

62 Rigging a Climate "Consensus": About those emails and "peer review," Wall Street Journal, Nov. 27, 2009.

63 The Global Warming Petition Project claims the support of thousands of scientists of various types and degrees who do not believe that anthropogenic climate change poses a serious risk to our planet within the foreseeable future. <www.petitionproject.org>. The website includes a paper summarizing some of the evidence in opposition to the reality of dangerous human-generated climate change.

64 The late-2009/early-2010 "Climategate" controversy has included criticism of some of the monitoring sites relied on by the United Nations. Some have suggested that there are problems that could have skewed the data and rendered the conclusions misleading, at least in some instances. *See, e.g.,* <http://www.timesonline.co.uk/tol/news/environment/article7026317.ece>.

65 I can testify from frigid personal experience that the coldest spot on the entire planet is a seat in the shadowy lower grandstand at Wrigley Field when the Cubs are playing there in early April and the wind is blowing in with gale force from the nearby lake. I don't care what the physicists may claim. Absolute zero is a temperature reading only possible at this location.

66 Al Gore, EARTH IN THE BALANCE: ECOLOGY AND THE HUMAN SPIRIT, Houghton Mifflin (1992).

67 THE CHRONICLE OF HENRY OF HUNTINGDON (T. Forester, editor and translator), Llanerch (1991), p. 199.

CHAPTER 3

1 John Charles Kunich, ARK OF THE BROKEN COVENANT: PROTECTING THE WORLD'S BIODIVERSITY HOTSPOTS, Praeger (2003).

2 John Charles Kunich, KILLING OUR OCEANS: DEALING WITH THE MASS EXTINCTION OF MARINE LIFE, Praeger (2006).

3 John Charles Kunich, *World Heritage in Danger in the Hotspots*, 78 IND. L. J.619 (2003); John Charles Kunich, *Preserving the Womb of the Unknown Species With Hotspots Legislation*, 52 HAST. L. J. 1149 (2001); John Charles Kunich, Fiddling Around While the Hotspots Burn Out, 14 GEORGETOWN INT'L ENVT'L L. REV. 179 (2001).

4 John Charles Kunich, *Losing Nemo: The Mass Extinction Now Threatening the World's Ocean Hotspots*, 30 COLUMBIA J. ENVT'L L. 1 (2005).

5 *See generally* Peter D. Ward, UNDER A GREEN SKY: GLOBAL WARMING, AND THE MASS EXTINCTIONS OF THE PAST, AND WHAT THEY CAN TELL US ABOUT OUR FUTURE (2007).

6 Nigel E. Stork, *The Magnitude of Global Biodiversity and its Decline*, in THE LIVING PLANET IN CRISIS: BIODIVERSITY SCIENCE AND POLICY, at 25 (Joel Cracraft & Francesca T. Grifo, eds. 1999).

7 IUCN Red List of Threatened Species <http://www.iucnredlist.org/>.

8   *See, e.g.*, Norman Myers, THE SINKING ARK: A NEW LOOK AT THE PROBLEM OF DISAPPEARING SPECIES (1979); THE GLOBAL 2000 REPORT TO THE PRESIDENT: ENTERING THE TWENTY-FIRST CENTURY 37 (1980); Paul Ehrlich, *Extinction: What Is Happening Now and What Needs to be Done*, in DYNAMICS OF EXTINCTION, 157 (David Elliott ed. 1986); Edward O. Wilson, *Vanishing Before Our Eyes*, TIME, April-May 2000, at 29-30; Nigel E. Stork, *The Magnitude of Global Biodiversity and its Decline*, in THE LIVING PLANET IN CRISIS: BIODIVERSITY SCIENCE AND POLICY, at 28 (Joel Cracraft & Francesca T. Grifo, eds. 1999).
9   *See* IUCN, Wildlife in a Changing World (2009) <http://data.iucn.org/dbtw-wpd/edocs/RL-2009-001.pdf>
10 *See* IUCN, Extinction crisis continues apace (3 Nov. 2009), <http://www.iucn.org/about/work/programmes/species/red_list/?4143/Extinction-crisis-continues-apace>
11 *See* S. Butchart, A. Stattersfield, and T. Brooks, Going or gone: defining 'Possibly Extinct' species to give a truer picture of recent extinctions, BULL B.O.C. 126A (2006) <http://www.birdlife.org/news/news/2006/06/possibly_extinct_paper.pdf>
12 Stuart Pimm et al, The Future of Biodiversity, 269 SCIENCE 347 (1995) (estimating that we are currently losing species at a rate between 100 and 1,000 times faster than the historical background pace).
13 Norman Myers, *Threatened Biotas: "Hot Spots" in Tropical Forests*, 8 ENVIRON. 1-20 (1988); Norman Myers, *The Biodiversity Challenge: Expanded Hot-Spots Analysis*, 10 ENVIRON. 243-56 (1990).
14 Russell A. Mittermeier, NORMAN MYERS, & CRISTINA GOETTSCH MITTERMEIER, HOTSPOTS: EARTH'S BIOLOGICALLY RICHEST AND MOST ENDANGERED TERRESTRIAL ECOREGIONS at 29-31 (2000).
15 Norman Myers, Russell A. Mittermeier, et al, *Biodiversity Hotspots for Conservation Priorities*, 403 NATURE 853, 855 (Feb. 2000).
16 *See generally* R. H. MacArthur, and E. O. Wilson, THE THEORY OF ISLAND BIOGEOGRAPHY, Princeton, N.J.: Princeton University Press (1967); David Quammen, THE SONG OF THE DODO: ISLAND BIOGEOGRAPHY IN AN AGE OF EXTINCTIONS, Scribner (1997).
17 Mann, *Extinction: Are Ecologists Crying Wolf?*, 253 SCI. 736, 737 (1991).
18 Robert MacArthur and Edward O. Wilson, THE THEORY OF ISLAND BIOGEOGRAPHY (Princeton University Press, 1967).
19 For some recent estimates, *see, e.g.*, Robert M. May, *The Dimensions of Life on Earth*, in NATURE AND HUMAN SOCIETY: THE QUEST FOR A SUSTAINABLE WORLD, 30-45 (Peter Raven, ed., NRC 1997) (estimating 7 million species worldwide, with a range from 5 to 15 million plausible); Nigel E. Stork, *The Magnitude of Global Biodiversity and its Decline*, in THE LIVING PLANET IN CRISIS: BIODIVERSITY SCIENCE AND POLICY, at 10-21 (Joel Cracraft & Francesca T. Grifo, eds. 1999) (hereinafter LIVING PLANET)

(employing various factors, taxon by taxon, in arriving at a rough estimate of 13.4 million species); Paul Williams, Kevin Gaston, and Chris Humphries, *Mapping biodiversity value worldwide: combining higher-taxon richness from different groups*, 264 PROCEEDINGS OF THE ROYAL SOCIETY, BIOLOGICAL SCIENCES 141-148 (1997) (crediting an estimate of 13.5 million species); Christopher Humphries, Paul Williams, and Richard Vane-Wright, *Measuring biodiversity value for conservation*, 26 ANNU. REV. ECOL. SYST. 93, 94-5 (1995) (accepting a range of 5-15 million species); T. Erwin, *Tropical forests: their richness in Coleoptera and other arthropod species*, 36 COLEOPT. BULL. 74-75 (1982); P.M. Hammond, *The current magnitude of biodiversity*, in GLOBAL BIODIVERSITY ASSESSMENT, 113-28 (V.H. Heywood, ed., Cambridge, 1995) (estimating 12 million species total); Norman Myers, *Questions of mass extinction*, 2 BIODIVERS. & CONSERV. 2-17 (1993) (mentioning several estimates and concluding that we can be fairly certain that there are at least 10 million species today).

20 Philippe Rekacewicz, Emmanuelle Bournay, UNEP/GRID-Arendal, Millennium Ecosystem Assessment <http://maps.grida.no/go/graphic/species-extinction-rates>

21 UNAIDS, World Health Organization, "2007 AIDS Epidemic Update," (Dec., 2007).

22 *See* Juliet Eilperin, *Wave of Marine Species Extinctions Feared*, Washington Post, August 24, 2005, at A1 (summarizing the strong evidence that a marine mass extinction is now underway, yet unnoticed by most people, because it is a "slow-motion disaster," both "silent and invisible").

23 David Jablonski, *Survival without recovery after mass extinction*, 99 PNAS (12), 8139-44 (2002); Chris D. Thomas, *et al. Extinction risk from climate change*, NATURE 427, 145-48 (8 January 2004).

24 *See* V.H. Heywood et al., *Uncertainties in Extinction Rates*, 368 NATURE 105 (1994).

25 *See* James Owen, Protected Areas Don't Protect Many Endangered Species, Study Finds, NATIONAL GEOGRAPHIC NEWS (7 April 2004), <http://news.nationalgeographic.com/news/2004/04/0407_040407_endangeredanimals.html>

CHAPTER 4

1 *See generally* Joyce McPherson, A PIECE OF THE MOUNTAIN: THE STORY OF BLAISE PASCAL, Greenleaf (1997).

2 O'Connell, Marvin R., BLAISE PASCAL: REASONS OF THE HEART, 29, (William B. Eerdmans Publishing Co. 1997).

3 *Id.* at 47.

4 *Id.* at 103.

5 *Id.* at 57.

6   Ross, John F., PASCAL'S LEGACY, European Molecular Biology Organization, Vol. 5, 58 (2004).
7   *See generally* Jeff Jordan, PASCAL'S WAGER: PRAGMATIC ARGUMENTS AND BELIEF IN GOD, Oxford University Press (2007); William G. Lycan and George N. Schlesinger, *You Bet Your Life: Pascal's Wager Defended*, in CONTEMPORARY PERSPECTIVES ON RELIGIOUS EPISTEMOLOGY, eds. R. Douglas Geivett and Brendan Sweetman, Oxford University Press (1992); Jeff Jordan, ed., GAMBLING ON GOD, Rowman & Littlefield (1994); James A. Connor, PASCAL'S WAGER: THE MAN WHO PLAYED DICE WITH GOD, HarperOne (2006).
8   *See generally* Alan Hájek, *Waging War on Pascal's Wager*, 112 PHILOSOPHICAL REVIEW 27-56 (Jan. 2003) for an excellent overview of the main arguments against Pascal's Wager in its traditional (religious) form. *See also* Nicholas Rescher, PASCAL'S WAGER, University of Notre Dame Press (1985).
9   Roy Sorensen, *Infinite decision theory,* in Jeff Jordan, GAMBLING ON GOD, Rowman & Littlefield (1994).
10 Jeff Jordan, *The many-gods objection and Pascal's wager,* 31 INTERNATIONAL PHILOSOPHICAL QUARTERLY 309-17 (1991); Paul Saka, *Pascal's wager and the many gods objection,* 37 RELIGIOUS STUDIES 321-41 (2001).
11 *See generally* Nassim Nicholas Taleb, THE BLACK SWAN: THE IMPACT OF THE HIGHLY IMPROBABLE, Random House (2007).
12 *See generally* Yakov Ben-Haim, INFO-GAP DECISION THEORY: DECISIONS UNDER SEVERE UNCERTAINTY, Academic Press (2006).

CHAPTER 5

1   *See generally* John W. Dawson, Jr., LOGICAL DILEMMAS: THE LIFE AND WORK OF KURT GÖDEL, A.K. Peters (2005); John L. Casti and Werner DePauli, GÖDEL: A LIFE OF LOGIC, Perseus (2001).
2   Dawson, John W., Jr., LOGICAL DILEMMAS: THE LIFE AND WORK OF KURT GÖDEL, 6 (A K Peters, Ltd. 1997).
3   Franzen, Torkel, GODEL'S THEOREM: AN INCOMPLETE GUIDE TO ITS USE AND ABUSE, 2 (A K Peters, Ltd. 2005).
4   *Id.* at 2.
5   Yourgrau, Palle, A WORLD WITHOUT TIME: THE FORGOTTEN LEGACY OF GÖDEL AND EINSTEIN, 4 (Basic Books 2006).
6   *Id.*
7   *Id.* at 11-12.
8   *Id.* at 156-157
9   *Id.* at 157

10   Stephen Hawking, editor, GOD CREATED THE INTEGERS: THE MATHEMATICAL BREAKTHROUGHS THAT CHANGED HISTORY, 1089 et seq., Running Press, Philadelphia (2005).
11   For an English translation of Gödel's original papers, *see* Kurt Gödel, ON FORMALLY UNDECIDABLE PROPOSITIONS OF PRINCIPIA MATHEMATICA AND RELATED SYSTEMS, Dover (1992).

CHAPTER 6

1   Cassidy, David C., BEYOND UNCERTAINTY, HEISENBERG, QUANTUM PHYSICS, AND THE BOMB, 17 (Bellevue Literary Press 2009).
2   *Id*. at 19.
3   Lindley, David, UNCERTAINTY: EINSTEIN, BOHR, AND THE STRUGGLE FOR THE SOUL OF SCIENCE, 85-86 (Anchor Books 2008).
4   *Id*. at 105-106.
5   *Id*. at 273.
6   *Id*. at 280.
7   *Id*. at 295.
8   *Id*. at 383.
9   *Id*. at 410.
10 *See generally* David C. Cassidy, UNCERTAINTY: THE LIFE AND SCIENCE OF WERNER HEISENBERG, W.H. Freeman (1993); Werner Heisenberg, ACROSS THE FRONTIERS, Ox Bow Press (1990).
11 *See* David J. Griffiths, INTRODUCTION TO QUANTUM MECHANICS, Prentice Hall (1995). For his role in developing the field of quantum mechanics, Heisenberg was awarded the Nobel Prize in Physics in 1932.
12 Werner Heisenberg, Über den anschaulichen Inhalt der quantentheoretischen Kinematik und Mechanik, 43 *Zeitschrift für Physik*, 172-198 (1927), English translation: J. A. Wheeler and H. Zurek, QUANTUM THEORY AND MEASUREMENT, 62-84, Princeton Univ. Press (1983).
13 Cassidy, D. Certain of Uncertainty. *in* UNCERTAINTY: THE LIFE AND SCIENCE OF WERNER WEISENBERG, 226-246, W. H. Freeman, (1993).
14 For a simple interactive on-line illustration of the butterfly effect in action, see <http://www.exploratorium.edu/complexity/java/lorenz.html>.
15 I am referring to the 1977 novel, THE THORN BIRDS, by Colleen McCullough, and the very popular television mini-series based on the book that first aired on NBC in 1983. It tells the heartbreaking, gut-wrenching tale of forbidden love and lust between an ambitious Catholic priest (bound by a vow of celibacy) and a much younger (and eventually married) woman. The main characters know that they must not yield to their temptations, but they ultimately do succumb, with far-reaching consequences.

16   Here is a partial list of logical fallacies. There are many on-line sources for such lists, often complete with examples and explanations. Some of the most common errors in logical reasoning include: (1) appeal to authority; (2) personal or ad hominem attack; (3) false dichotomy; (4) negative proof fallacy, that because we can't prove something false it must be true, or vice versa; (5) red herring distractions; (6) appeal to probability, that because something could happen it necessarily will happen; (7) argument from repetition; (8) reductio ad absurdum, making an argument appear ridiculous by exaggerating it or its conclusions; (9) cherry picking evidence or examples selectively and unfairly; (9) post hoc ergo propter hoc, that an earlier event must have caused a later one; (10) appeal to majority opinion; (11) hasty generalization; (12) appeal to force or coercion; (13) guilt by association; (14) appeal to emotion (fear, flattery, wishful thinking, etc.); (15) attacking an argument by questioning the motives of its advocate; (16) appeal to tradition; (17) ignoring exceptions to a generalization; (18) wrong direction, where cause and effect are reversed; (19) incomplete or inconsistent comparison; (20) causal oversimplification; and numerous others.

CHAPTER 7

1   This is from the delineation contained in the Rio Declaration on Environment and Development of 1992, Principle 15. Rio Declaration on Environment and Development, June 14, 1992, UN Doc. A/CONF.151/5/Rev.1 (1992), available at: <http://www.unep.org/Documents.multilingual/Default.asp?DocumentID=78&BookID=1163>. *But see* Catherine Tinker, *Is a United Nations Convention the Most Appropriate Means to Pursue the Goal of Biological Diversity?: Responsibility for Biological Diversity Conservation Under International Law*, 28 VAND. J. TRANSNAT'L L. 777, 779 (Oct. 1995) (stating that there is no agreement on the content of the Precautionary Principle).
2   *See* Arie Trouwborst, EVOLUTION AND STATUS OF THE PRECAUTIONARY PRINCIPLE IN INTERNATIONAL LAW (The Hague, London, Boston: Kluwer Law International) (2002) (concluding that "the Precautionary Principle is not only a general, perhaps even universal custom in that it binds, in principle, all governments of the world, but also in that it aims for comprehensive environmental protection.") *Id.* at 284. *See also* WTO Appellate Body Report on EC Measures Concerning Meat and Meat Products (Hormones), WT/DS26/AB/R, WT/DS48/AB/R, AB-1997-4 at para. 121 (Jan. 16, 1998) (stating that the Precautionary Principle "at least outside the field of international environmental law, still awaits authoritative formulation" as customary international law).
3   *See, e.g.*, United Nations Framework Convention on Climate Change (FCCC), May 9, 1992, 31 ILM 849 (1992); Convention on Biological Diversity, June 5, 1992, 31 ILM 818 (1992); Kyoto Protocol to the FCCC, Dec. 10, 1997, 37 ILM 22 (1998); Cartagena Protocol on Biosafety to the Convention on Biological Diversity, Jan. 29, 2000, 39 ILM 1027 (2000).

4   *See* Stewart, Terence P. Stewart, David S. Johanson, *A Nexus of Trade and the Environment: The Relationship Between the Cartagena Protocol on Biosafety and the SPS Agreement of the World Trade Organization*, 14 COLO. J. INT'L ENVTL. L. & POL'Y 1, 40, at 40-44 (Winter 2003). (arguing that the Precautionary Principle may not have attained customary international law status due primarily to the lack of a sufficient time lapse, and pointing out that none of the major international tribunals have yet ruled on this point.) *See also* Catherine Tinker, *Is a United Nations Convention the Most Appropriate Means to Pursue the Goal of Biological Diversity?: Responsibility for Biological Diversity Conservation Under International Law*, 28 VAND. J. TRANSNAT'L L. 777, 795 (Oct. 1995) (arguing that the Precautionary Principle may not have attained customary international law status due to the fact that the international instruments which have included this principle are neither binding nor intended to be binding upon the parties); John D. Graham, The Role of Precaution in Risk Assessment and Management: An American's View, Address Before the European Commission, the U.S. Mission to the E.U., the German Marshall Fund with the European Policy Centre and the Center for Environmental Solutions (Jan. 11-12, 2002).

5   *See generally* Wouter de Been, LEGAL REALISM REGAINED: SAVING REALISM FROM CRITICAL ACCLAIM, Stanford Law (2008).

6   *See generally* William W. Fisher *et al.*, AMERICAN LEGAL REALISM, Oxford University Press (1993); John Henry Schlegel, AMERICAN LEGAL REALISM AND EMPIRICAL SOCIAL SCIENCE, University of North Carolina Press (1995).

7   *See generally* Roberto Mangabeira Unger, THE CRITICAL LEGAL STUDIES MOVEMENT, Harvard University Press (1986); Richard W. Bauman, CRITICAL LEGAL STUDIES: A GUIDE TO THE LITERATURE, Westview (1996).

8   John Charles Kunich, ARK OF THE BROKEN COVENANT: PROTECTING THE WORLD'S BIODIVERSITY HOTSPOTS, 177-84, Praeger (2003).

9   John Charles Kunich, KILLING OUR OCEANS: DEALING WITH THE MASS EXTINCTION OF MARINE LIFE, 138-53, Praeger (2006).

10 Kunich, note 8, at 179.

11 *Id* at 149-151.

12 *Id* at 13-18.

13 *See* Christopher Booker, "Climate Change: This is the worst scientific scandal of our generation," Telegraph, Nov. 28, 2009. <http://www.telegraph.co.uk/comment/columnists/christopherbooker/6679082/Climate-change-this-is-the-worst-scientific-scandal-of-our-generation.html>

CHAPTER 10

1. John Charles Kunich, KILLING OUR OCEANS: DEALING WITH THE MASS EXTINCTION OF MARINE LIFE, at 143-51 Praeger (2006). My apologies to the many fans (including my daughters) of the incredibly popular Disney Channel Original Movie *High School Musical*, which includes a song entitled, "Stick to the Status Quo."
2. John Charles Kunich, ARK OF THE BROKEN COVENANT: PROTECTING THE WORLD'S BIODIVERSITY HOTSPOTS at 182, Praeger (2003).
3. *See* Thomas Lovejoy, *Species Leave the Ark One by One*, in THE PRESERVATION OF SPECIES: THE VALUE OF BIOLOGICAL DIVERSITY 16-18 (Bryan Norton ed. 1986). The Penicillium mold is a classic example of unexpectedly vast utility from a living entity. Useless or an annoyance for thousands of years, this inconspicuous, humble mold was then found to ward off competitive fungi, which made it useful in producing and preserving Roquefort cheese. This use was then the foundation for the profound antibiotic medical advancements that have in large part catapulted the human race out of the era of early death. Similarly, aspirin, or salicylic acid, consists of an organic molecule originally derived from a willow, Salix. *Id. See also* William Barrett, *Delaying Tactics*, FORBES, Mar. 1998, at 68 (stating that the Pacific yew plant, which was once burned as a pest in the Northwest old growth forests, was discovered to be the source of the anti-cancer drug Taxol and now brings the Bristol-Meyers Squibb Company in excess of $1 billion per year).
4. John Charles Kunich, KILLING OUR OCEANS: DEALING WITH THE MASS EXTINCTION OF MARINE LIFE, 148-49, Praeger (2006).
5. *See generally* William D. Nordhaus, A QUESTION OF BALANCE: WEIGHING THE OPTIONS ON GLOBAL WARMING POLICIES, Yale University Press (2008) for a thoughtful economic analysis of the possible costs and benefits of climate change intervention under a fairly wide spectrum of scenarios.
6. *See* Christopher Booker, "Climate Change: This is the worst scientific scandal of our generation," Telegraph, Nov. 28, 2009. <http://www.telegraph.co.uk/comment/columnists/christopherbooker/6679082/Climate-change-this-is-the-worst-scientific-scandal-of-our-generation.html>
7. *See generally* Martin J. Rees, GLOBAL CATASTROPHIC RISKS, Oxford University Press (2008) for several examples of potentially disastrous events and approaches we might take to planning for such unlikely but devastating contingencies.
8. *Id.*

INDEX

– A –

Abbott, Edwin 160
Abby Sunstein 48
Abyss 99
Accidents 82, 105, 108, 123, 239-240, 247
Acid rain 19
Adjustment 30, 80, 120, 132, 137, 182-183, 224-225, 259, 300, 322
Agriculture 110, 123, 219
AIDS 12, 109-110
Air pollution 79, 207
Air travel 250-253
Airplanes 250-253
Alabama, University of 47
Alarmism 24, 63, 66
Alcohol 135, 193, 195, 239, 242, 251, 276, 282
All-in wager 225, 289, 293, 312-313, 323, 342
Alternatives analysis 132, 139, 141, 145, 149-150, 185, 233-234, 241, 243-244, 249-250, 252-253, 271-272, 327, 348
Amazonia 89, 96
Annelids 95, 98, 103, 112
Antarctica 33, 42, 75
Anthropogenic climate change 46-49, 54-55, 57-58, 60-61, 66, 219
Antibiotics 107-109
Apathy 244, 315
Archaea 98-99, 103
Arithmetic 158
Ark of the Broken Covenant 86, 207
Asbestos 19, 205
Assumptions 13, 56, 73, 96, 100, 118, 132-133, 137, 146-147, 149, 158, 186, 190, 194, 242, 254, 261, 274, 291, 298, 313-314, 329-330, 351
Atmosphere 32-33, 36-37, 39, 43, 46, 51, 54, 56, 78-80, 113, 115, 181, 219, 316
Atomic weapons 174-175
Autism 255-256, 350
Axioms 153, 155-156

– B –

Baccarat 278, 286
Background extinction rate 88
Bacteria 98, 107, 109
Bad beat 136, 328
Bad luck 136, 328
Balance of trade 318
Bald Eagle 19
Ball, Timothy 57
Beetles 97, 106, 112
Bias 23, 157, 162, 199, 213, 217, 311, 328, 351
Bible 27
Bingo 351, 356
Biodiversity 86-93, 99, 107, 115, 118-121
Biodiversity hotspots 90-93, 115, 118, 208-210, 314-315
Blackjack 279, 285-286
Black swan theory 145-149, 339
Bluff 289-290, 295, 303
Bohr, Neils 172, 175, 236
Bomb, atomic 174-175
Book of Mormon 27
Brazil 62, 94
Burden of proof 200-205, 333, 343
Bureaucracy 29
Burns, Robert 237
Butterfly effect 182-183, 255

– C –

Cancer 12, 228, 245-246, 330
Cap and trade 54, 79, 317-318
Carbon 32, 317
Carbon dioxide 32-33, 39, 45, 51, 54, 80, 219
Carbon fixation 33, 113, 219
Card games 279
Carlin, Alan 57-58
Carson, Rachel 18-19
Casino gambling 274-285
Cause or contribute finding 50-51
Caution 233-234, 296-298
Celebrities 309, 314
CERCLA 19
Chaos theory 56, 182-184

— 371 —

Charybdis 298
Chicago 70-71
Chicago Cubs 146, 322
Chickenpox 254
China 43, 50, 62, 79
Christie Kunich 48, 159
Cigarettes 21, 244-249
CITES 118
Civil trials 201-206, 223
Clean Air Act 19, 50-52, 79, 325
Climategate 11-12, 63-67, 217, 320, 324
Climate change 29, 31-84, 125-126, 213-215, 218, 229, 299, 305, 307, 313-315, 347
Climate change decision matrix 213-215, 220-223, 313-315, 326, 340-341
Climate wager 214-223, 318-319, 340
Climatic Research Unit (CRU) 60, 63-65, 217, 320
Clinton, Hilary 20
Clouds 39, 44, 316
Coal 318
Coastal zones 43
Committed to extinction 115-116
Competition 38, 116, 214, 317
Compulsive gambling 272, 354
Congress 18-19, 52
Consensus 22, 53, 58-59, 64, 66, 74
Conservation 118-119, 122-123, 315
Consistent systems 156
Coral reefs 93
Cost/benefit analysis 61, 77-78, 122-123, 222-223, 247-248, 319-320
Craps 279, 282, 351
Criminal trials 200-206, 223, 330, 332
Crisis 17-18, 20
Critical Legal Studies 198-199
Crops 110. 219
Crustaceans 106

– D –

DDT 19
Decision analysis 127, 129, 133, 184, 186, 194, 297, 327
Decision-making 25-26, 104, 133, 136, 140, 184, 186-188, 192-193, 195-196, 200, 206, 209, 227, 232-234, 237-238, 270, 299, 307, 326-327, 341, 349-350
Decision matrix 145, 186, 210-212, 218, 220, 222-224, 247, 310-314, 326, 340-343, 347
Decision theory 131, 140-152, 184-185, 327
Decisions under risk 144
Decisions under uncertainty 137, 147-148
Decomposition of waste 112-113
Default option 15-16
Deforestation 81-82, 315
Denialists 23
Desertification 44
Development 23, 123, 317
Diet 106, 246, 264
Dilemmas 13, 293, 298-299, 309, 333, 356

Dimensions, three 160-162, 348
Dimensions, two 160-162, 348
Dinosaurs 85, 115
Diphtheria 254
Disease 107-108, 113
Disagreement 22, 28-29, 336
Disaster 187
DNA 110
Dodo 20, 93, 115, 258
Dogma 24, 214, 225
Doubt 13, 22, 30, 95-96, 267, 274, 294, 307-308, 310, 354
Driving 238-242
Drought 44, 47

– E –

Earth 14, 32, 316, 352, 355
East Anglia, University of 63
Ecosystems 45, 68, 89, 111, 120, 227
Ecosystem services 111-113, 311
Efficiency 104
Einstein, Albert 152-153, 174, 176, 180-181, 236
Electrons 179, 236
E-mails 63
Eminent domain 119
Emissions 33, 44, 51-52, 55, 58, 60-62, 79, 316-317
Emotion 186, 190-192, 259, 284, 328-329, 346
Endangerment determination 51
Endangered species 18-19, 85, 101, 104, 122, 227
Endangered Species Act 18-19, 118-119
Endemism 89, 120
Endemic species 89-90, 120
Energy 207, 319
Environmental law 18-19, 47, 84, 154, 168, 198, 308
Environmental Protection Agency (EPA) 50-51, 60, 325
Epidemic 109, 116, 310
Epistemological issues 180-181, 310
Error of the first kind 329-334
Error of the second kind 331-334
Error, Type I 329-334, 338
Error, Type II 331-334, 338
Error, Type R 335-344, 349, 351
Error, Type P 335-344, 349, 351
Evidentiary standards 200-202
Exercise 259-264
Expected Value (E.V.) 141-144, 244, 282
Expense 311-312, 319-320
Experts 22, 314, 353
Extinction 12, 84, 86-87, 114-116
Exotic species 315
Evidence 200-202
European Union 321

— F —

Faith 24-27
Fallacies 188-190
False negative 331-334
False positive 329-334
Feedback mechanisms 32, 39, 44
Fermi, Enrico 174
Fire insurance 345
Flatland 160-170, 348-353
Fleming, Alexander 108
Flooding 12, 37, 43, 69, 227
Flying 250-253
Food 45, 80, 89, 91, 94, 102, 104-107, 110-112, 273, 311
Fossil fuels 29, 39, 45, 54, 61, 78-80, 318
Framework Convention on Climate Change 62
Fuel 54, 80, 318
Fugitive emissions 33
Fungi 98-99

— G —

Galileo 66
Gambling 12, 124, 235, 250, 265, 266-305, 354-355
Gambling problem 300, 346-347, 354-355
Game theory 131, 140, 149
Gaps in information 11, 40, 126, 231-232, 247-248, 295-296, 304, 307, 324
Genetic resources 110, 123
Genetically modified organisms 29
Geothermal power 80
Germany 171
Glaciers 33, 42, 45
Global mean temperature 40-41, 46, 53, 75, 316
God 128, 131-138, 145
Gödel, Kurt 14, 151-155, 232, 309, 312, 314, 324-325, 327, 334-335, 340, 347-348
Gordian knot 306
Gore, Al 61, 79
Gradual threats 21, 75-76, 114-116, 225-226, 261-264, 302, 339
Greenhouse gasses 32-33, 39, 51, 54, 78, 80, 113, 317, 325
Greenland 33, 42, 56

— H —

Habitat destruction 14, 47, 120-121, 315
Habitat modification 14, 44, 47, 120-121
Hackers 63
Hannah Montana 344
Hazardous waste 14, 19
Heart disease 12, 245-246, 248-249
Heisenberg, Werner 14, 171-176, 232, 236, 309, 314, 325, 334-335, 344, 349
Heliocentrism 66
Hellmuth, Phil 290
Hepatitis 254
High stakes 12, 299, 310, 335, 354
Himmler, Heinrich 173-174
Hitler, Adolf 173, 175
Hockey stick graph 35-38, 72, 217
Horizon of uncertainty 147-148
Hotspots, biodiversity 90-93, 115, 118, 208-210, 314-315
Hotspots decision matrix 210-213, 311, 341
Hotspots wager 207-211, 222, 310-311
House percentage 278-279, 282, 285-286
Humidity 39
Hurricanes 47
Hydrocarbons 32-33, 39
Hydrothermal vents 89, 99

— I —

Ice cap 12, 33, 47
Ice samples 33
Ignorance 24, 100, 121, 162, 190, 349
Illogical decision-making 184-196, 347
Immediacy of threats 15, 17, 20, 77, 184
Immunization 253-259
Importance of biodiversity 101-114
Inaction 136, 200, 208, 225, 233-234, 243, 248-249, 252-253, 264-265, 272-273, 283, 296-298, 301-302, 308
Incompleteness theorems 155-158, 168-169, 348-349
Inconvenience 49, 102, 139, 157, 222, 227, 253, 337, 341
India 50, 62, 79
Industrial activity 58, 62, 81
Industrial Revolution 34, 36-37, 78
Industry 55, 58, 224, 318-319, 357
Inertia 233-234, 243, 248-249
Infinity 131-132, 136-137, 140, 145, 233
Info-gap 141, 147-149, 339
Inoculation 253-259, 350
Insects 45, 95-98, 106, 113
Insurance 195, 224-228, 312, 319-320, 344-345
International law 197-198
International Union for the Conservation of Nature (IUCN) 87-88
Intergovernmental Panel on Climate Change (IPCC) 35, 55-56, 59-60, 64
Intervention 21, 29, 31, 48, 61, 86, 100, 117, 139, 183, 209, 214, 216, 220, 222, 224, 231, 308, 311, 313-315, 319-322, 324, 337, 340-343
Invasive species 116, 315
Irrationality 25, 185, 192, 213, 233-235, 282, 313
Irresponsible behavior 270, 282, 354
Islands 43, 90-92, 115
Island biogeography 90-92
IUCN 87-88

— J —

Jackpot 266-274
Jeopardy 323, 342
Julie-Kate Kunich 48, 159

— K —

Kepler, Johannes 66
Killing Our Oceans 86, 207
King Canute 81-82
Kunich, Christie 48, 159
Kunich, Julie-Kate 48, 159
Kunich, Mae 130
Kyoto Protocol 50

— L —

Lake Chad 44
Land use 319
Las Vegas 265, 285
Law students 200, 235
Leaders 50, 62, 162, 208-209, 236, 303
League of Nations 317
Legal Realism 198-199
Lifestyle choices 228, 245, 253
Life insurance 226-228, 233, 273, 319-320, 344
Lineland 161-162, 166-167
Litigation 53, 205, 326
Little Ice Age 36-37
Livestock 33, 79-80, 104, 110, 112
Living dead 115-116
Logic 48, 141, 144, 153-154, 158, 169, 180, 186-188
Logical fallacies 188-190
Long-term dangers 16, 45, 194, 225-226, 261, 339
Long-range perspective 193-194
Lorenz, Edward 56, 181-182
Lotteries 266-274, 350
Love Canal 19
Luck 253, 269, 286-287, 290, 345-346, 356
Ludic fallacy 146-147

— M —

Madagascar 94
Main Event, World Series of Poker 285, 292-293, 313
Manhattan Declaration 58
Mann, Michael 35
Mass extinction 12, 29, 84, 86, 92, 114-117, 125-126, 208-210, 222, 228, 299, 305, 307, 310-311, 347
Massachusetts v. EPA decision 50-53, 61, 325-326
Mathematics 129, 151-152, 154-158, 172, 199, 296, 317
Maximin 149
McDuck, Scrooge 274
Measles 254

Medical examination 138-139
Medicine 107-108, 111, 262, 273, 311
Medieval warm period 36-37, 55, 72-73
Mekong Delta 43
Melting, ice cap 47
Memorial (Pascal) 128-129
Methane 32-33, 45, 51, 80
Minimax 141, 149-150, 296-297
Misconceptions 76
Mites 95, 98
Monera 99
Money Bin 274
Monitoring stations 40, 68
Moral reason to save biodiversity 101-103
Mountains 42, 69
Mumps 254
Murphy's Law 195-196, 223
Myths 115, 356-357

— N —

National Academies of Sciences 38
National Aeronautics and Space Administration (NASA) 41, 60
National Oceanic and Atmospheric Agency (NOAA) 36, 60
National Snow and Ice Data Center 57
National sovereignty 321
Nature 121, 315, 356-357
Nazis 173-175
Negative feedback effect 44
Nematodes 95, 98, 103
Newton, Isaac 180
Newtonian physics 181
Neyman, Jerzy 327, 329, 335
Night of fire (Pascal) 128
Nitrogen oxides 51
No limit Texas Hold 'em 285-305
Nordhaus, William 61
Nuclear energy 14, 29, 80, 318
Null hypothesis 329-334

— O —

Obama, Barack 20
Obesity 21, 260, 263
Objectivity 233-234, 351
Observer effect 176
Oceans 33, 69, 75, 98
Ocean currents 47, 54, 316
Ocean levels 43, 73, 217
Odds 50, 143-144, 191-193, 225, 232, 237, 242, 244, 248, 258, 265, 267, 274-276, 288, 297, 302, 307-308, 356
On tilt 289, 328-329
Opinion, public 119
Opportuneness 148-149
Opportunity costs 78, 122, 224-225, 243, 252-253, 319-320, 338-339, 341
Orbit, fluctuations in 47, 54
Orwell, George 234

Outliers 146
Outside model occurrences 145
Overcautiousness 233-234, 296-298, 301-302, 339
Overfishing 105-106
Ozone layer 19, 28-29, 207

– P –

Pandemic 109-110, 226
Pandora 326, 357
Papua New Guinea 94
Pascal, Blaise 14, 127-138, 168-169, 232, 309, 314, 323-324, 327, 334-335, 340, 343, 349
Pascal's Wager 129-142, 168-169, 207, 244, 312, 320, 345
Passivity 243-244, 301, 335-337
PCBs 19
Pearson, Egon 327, 329, 335
Peer review 55, 63-64
Penicillin 107-108, 110, 211
Pensées 130-132
Perfluorocarbons 51
Pertussis 254
Petroleum 318
Pesticides 19
Pets 80, 103-104, 106
Photon 177
Photosynthesis 33, 39, 80, 113, 219
Physics 55, 152, 172-173, 175-176, 180-181, 317
Physical examination 138-139, 263
Plague 109-110
Plants 33, 39, 70, 80, 97-98, 104, 106, 112-113, 219, 310
Plimer, Ian 57
Poaching 122, 315
Poker 13, 249, 285-305, 323, 328, 338, 342, 351, 354, 356
Polar ice cap 33, 47
Policy 14, 16, 18, 21, 24, 26-27, 47, 204, 207, 232
Polio 254
Politics 15, 20, 74, 308, 317, 326
Politicians 15, 20, 303, 308-309, 326
Pollination 112
Pollution, Air 79, 207
Pollution, Transboundary 14, 29, 207
Pollution, Water 21, 105
Population problems 14, 207
Positive feedback effect 44
Poverty 271
Precautionary Principle 197-199, 222, 231, 308, 345, 347
Precipitation 43-44, 112, 251
Pregnancy 331
Prejudice 157, 186, 199, 313
Preponderance of the evidence 202-205
Presumption of innocence 200, 204, 330, 332-333
Probabilities 12, 22, 41, 49, 133, 141, 145, 147, 150, 169, 185, 190, 211, 225, 232-233, 236, 242, 245, 280-282, 300, 304, 323
Probability distribution 236
Prometheus 356-357
Prostate cancer 330-331
Protected habitats 120, 122
Proof 25, 28, 48, 75, 155, 160, 166, 168, 200-207, 223, 231, 333-334, 342-344, 348
Proof beyond a reasonable doubt 200, 203
Protista 99
Provability 155-156, 169, 310, 347, 349-350

– Q –

Quality of life 61, 78, 206, 222, 246, 253, 256, 262, 299, 337, 340
Quantum mechanics 176-177, 181-182, 300, 314
Quran 27

– R –

Rainforests 69, 89, 93, 97-98, 123
Reasons for protecting biodiversity 101-114
Recklessness 82, 281-282, 302, 312, 355
Red list 87-88
Regulated pollutant 51-52, 326
Relativity, Theory of 173, 176
Remote Sensing System 47
Reward 12, 25, 41, 82, 131-141, 169, 222, 234, 237, 242-243, 264, 267, 273-274, 282, 296, 300, 303, 323, 337, 339, 342, 344, 346, 356
Risk 11-12, 25, 41, 82-83, 99-104, 118, 129, 131, 136, 140-141, 144-146, 149, 169, 187, 191, 193, 201, 206, 211, 219, 222-225, 228, 233-235, 237, 240-245, 248-250, 257-258, 272-275, 283, 294-302, 312
Risk tolerance 241, 258-259, 295, 299
Robustness 148-149
Roulette 142, 278, 286
Rubella 254
Ruin 90, 204-205. 224, 259, 295, 297-298, 302, 312, 339, 356

– S –

Safety 233-234, 239-241, 253
Satisficing 148
Scandal 64-67, 217, 320, 324
Schroedinger's cat 331
Scientific disagreement 53, 63-66
Scientific method 63, 65
Scylla 298
Seafood 105-106
Sea level 43, 73, 217
Silent Spring 18-19
Skepticism 44, 50, 258, 331
Skill 279, 286-287, 346, 350
Slot machines 276-277, 279
Slow-acting hazards 16, 302, 340
Smoking 21, 244-250
Social Security 16

Societal changes 61
Solar activity 47
Solar power 80
Solar System 66
Sovereignty, national 321-322
Spaceland 161-162, 166-167, 348
Species, endangered 18-19, 85, 101, 104, 122, 227
Species, number of 89, 93, 95, 99
Species, undiscovered 94-95, 99, 118
Spiders 98
Stark, Johannes 173
Statistical significance 330-331
Status quo 14-15, 22, 248-249, 343
Sulfur hexafluoride 51
Sun 47, 54
Sunstein, Abby 48
Superfund 19
Supreme Court 50, 325-326
Sustainability 118-119, 122-123, 315
Symbiosis 111

— T —

Tailoring rule 52, 325
Taxes 16, 58, 204, 236
Technology 37, 52, 56, 69, 316
Tells, in Poker 294
Temperature, global 40-41, 46, 53, 75, 316
Tetanus 254
Theory of Relativity 173, 176
Third Reich 173
Threatened species 87, 119, 315
Three Mile Island 21
Ticks 98
Tilt 289
Tobacco 21, 244-250
Toxic waste 19
Transboundary pollution 29, 207
Transportation 243, 251, 319
Tree rings 34-35, 75
Truth 25-26, 58, 64, 66-68, 135, 155, 157-160, 165, 169, 180, 199, 237, 267, 311, 324, 347-348, 353
TSCA 19
Twain, Mark 236-237
Type I errors 329-334, 338
Type II errors 331-334, 338
Type P errors 273, 283, 335-344, 340, 351
Type R errors 273, 283, 335-344, 349, 351

— U —

Uncertainty 13-14, 22, 29, 76-77, 82, 89, 99-100, 126, 180, 231-232, 237, 245, 295-296, 307, 310, 334, 353, 357
Uncertainty principle 171, 176-181, 183
Uncertainty, severe 147-148
Unknowns 12, 94, 99, 136, 251-252, 256-257, 262-263, 271-274, 280-282, 290-292, 304, 307

Uncle Scrooge 272
Unemployment 205, 222, 318
United Nations 38, 317, 321
United States 61, 79, 208, 317-318
Unused insurance 224, 312, 319, 323, 341
Utilitarianism 104
University of Alabama 47
University of East Anglia 63

— V —

Value of biodiversity 104-110
Variables 214-216, 235, 240-241, 246-247, 251-252, 256-257, 262-263, 271-274, 280-282, 290-292, 303-304
Vegetation 33, 39, 80, 90, 92,
VFS (Voluntary Flatland Syndrome) 350-353
Viruses 109
Volcanic activity 54
Voluntary Flatland Syndrome (VFS) 350-353

— W —

Warming, global 11, 31-32, 47, 53, 55-57, 59-61, 64, 66, 72, 76, 83, 87, 207, 213, 318, 325, 357
Warming trend 47, 59
Wasted resources 312, 319
Water pollution 21, 105
Water vapor 39, 55-56
Weather 37, 44, 47, 70, 75, 181-183, 239-240, 251, 253
Weather patterns 43
Weather, severe 43, 47
Wilderness preserves 118-120
Wildlife refuges 118-120
Wind power 80
Windfalling 148
World decision matrix 210-215, 248, 310, 326, 347, 351-352
World government 321-322
World Series of Poker 285, 292-293, 313
World Wager 207-218, 232, 238, 248, 310, 313, 326, 344-345, 347, 351-352
World War II 174-175
Worst-case scenario 149-150, 210, 220, 233, 296-297, 311, 345
Wrigley Field 182, 242

— Y —

Yangtze Delta 43

— Z —

Zeus 356-357